COOPERATIVE INNOVATION

Science and Technology Policy

World Scientific Series in R&D Management

Print ISSN: 2591-7498
Online ISSN: 2591-7501

Series Editor: Tugrul U Daim *(Portland State University, USA)*

Published

Vol. 3 *Cooperative Innovation: Science and Technology Policy*
by Frederick Betz

Vol. 2 *Technology Roadmapping*
edited by Tugrul U Daim, Terry Oliver and Rob Phaal

Vol. 1 *Managing Technological Innovation: Tools and Methods*
edited by Tugrul U Daim

World Scientific Series in R&D Management – Vol.3

COOPERATIVE INNOVATION

Science and Technology Policy

Frederick Betz

Portland State University, USA

World Scientific

EW JERSEY • LONDON • SINGAPORE • BEIJING • SHANGHAI • HONG KONG • TAIPEI • CHENNAI • TOKYO

Published by

World Scientific Publishing Co. Pte. Ltd.
5 Toh Tuck Link, Singapore 596224
USA office: 27 Warren Street, Suite 401-402, Hackensack, NJ 07601
UK office: 57 Shelton Street, Covent Garden, London WC2H 9HE

Library of Congress Cataloging-in-Publication Data
Names: Betz, Frederick, 1937– author.
Title: Cooperative innovation : science and technology policy / Frederick Betz
Description: New Jersey : World Scientific, [2019] | Series: World Scientific series in R&D management | Includes bibliographical references.
Identifiers: LCCN 2018046973 | ISBN 9789813238695 (alk. paper)
Subjects: LCSH: Technological innovations--Government policy. | Technology and state. | Science and state. | Research, Industrial--Government policy. | Research--Government policy. | Academic-industrial collaboration.
Classification: LCC HC79.T4 B48 2019 | DDC 338.9/26--dc23
LC record available at https://lccn.loc.gov/2018046973

British Library Cataloguing-in-Publication Data
A catalogue record for this book is available from the British Library.

For any available supplementary material, please visit
https://www.worldscientific.com/worldscibooks/10.1142/10951#t=suppl

Desk Editors: Nur Izdihar Binte Ismail/Amanda Yun

Typeset by Stallion Press
Email: enquiries@stallionpress.com

Preface

To fully understand why some nations are more successful in innovation than others, one must examine the structure and process of knowledge creation and use — the science & technology policy of a nation. Knowledge of nature is discovered by science, and knowledge of the manipulation of nature is invented by technology. Technology is embedded into commercial products/services by industrial design. National innovation requires progress both in science and technology and also in economy.

Research cooperation for innovation is necessary, since science, technology, and production are performed in different sectors of a nation. Universities perform most science, and science discovers nature. Government supports most of the research in universities, and therefore is the principal sponsor of science. Industry develops most technology and commercializes technology in economically useful products/services. The structure and process of knowledge in a nation thus requires (1) the creation of knowledge in science, (2) the translation of science into technology, and (3) the design of technology into commercialization of utility.

National innovation consists of discovery, invention, and commercialization. Research cooperation is necessary for a competitive science & technology policy.

Macro-innovation, innovation at the national level, is a complicated picture — depicting the proper ways that government-university-industry can cooperate to advance knowledge and economically benefit from innovation. We will review the special programs in science & technology policy that have proven beneficial in fostering research cooperation for national competitiveness, as depicted in the following figure.

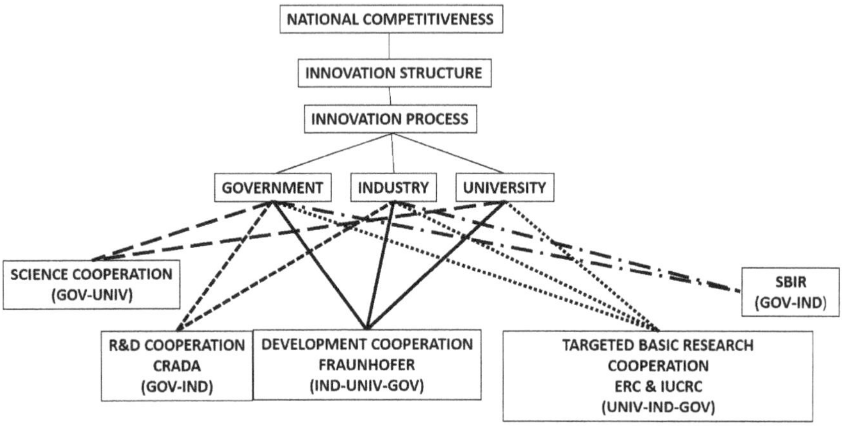

Figure. Types of Cooperative Research Programs of Government S&T Policy.

In this book, we will review these programs in detail, with historical case studies.

Science Cooperation Program — to ensure government funding of university science for national scientific excellence.

R&D Cooperation CRADA Program — to facilitate science and technology transfer from government science into industrial technology.

Development Cooperation Fraunhofer Program — to facilitate university research assistance to innovative product development (including small businesses).

Targeted-Basic Research Cooperation Program — to facilitate university and industrial research cooperation toward Next Generation Technologies.

Small Business Innovation Research — to provide government research support to small businesses in the first stages of innovation.

About the Author

Frederick Betz is occasionally Adjunct Professor in the Department of Management of Engineering and Technology, Portland State University, Portland, Oregon, USA. He received a PhD in physics from the University of California at Berkeley, but subsequently changed fields to operations research. He has taught at graduate business schools in several universities and served as a program officer in the U.S. National Science Foundation. He has recently taught as an adjunct or visiting professor at Portland State University in Oregon, Nile University in Egypt and Korea University and SUNY Korea in South Korea. He has published in the fields of management, philosophy of science, management of technology, and political economics. His recent books in print include *Stability in International Finance — Application of Disequilibrium Theory* (2016), *Strategic Thinking — Vision, Competence, & Integrity* (2015), *Modeling Methodology For Sustainable Development* (2015), *Innovation and Global Competition: The Case of Korea* (Co-authored with Youngrak Choi, Whasik Min, Won Koo Park, and Dong Woo Shin) (2014), *Why Bank Panics Matter* (2013), *Societal Dynamics* (2011), *Managing Science* (2010), and *Managing Technological Innovation (3rd Ed.)* (2010).

Contents

Chapter 1

Innovation and National Competitiveness

Introduction

Since the industrial revolution in the late 1700s, innovation has been the driving force behind economic growth and advancement in the military might of nations; namely, USA, UK, and a number of European countries. The emergence of science in the 1600s created the environment for the development of technology to power the world ahead for the next three centuries. Yet, one problem persists; very few nations have truly mastered the art to manage innovation, which is the manner to institutionalize the process to advance science and technology, to accelerate economic growth and empower the government.

The process of innovation begins with 'research' and moves to 'development'. This is called 'Research & Development' (R&D). Research discovers and explains nature, sometimes even inventing a new way to manipulate things. Using new technology, development creates new products or services to generate revenue. R&D develops, from science, technology and invents new products.

Take, for example, the discovery of DNA. As scientists, Watson and Crick discovered the geometric form of the gene, the double-stranded helix model, through scientific investigation. That inspired Boyer and Cohen (also as scientists) to discover a way to manipulate and alter genes through the technique of recombinant DNA. This paved the way for Boyer and Swanson to start the first biotech company, Genentech, to produce therapeutic proteins by manipulating the genes of e-Coli bacteria. From this (1) science of DNA, (2) technology of recombinant DNA, and (3) commercialization of therapeutic proteins, the new biotechnology industry was born. This has been the pattern of innovation: (1) scientific discoveries, (2) technological inventions, and (3) commercial ventures to produce new products using new technology. *Innovation is science combined with invention plus commercialization of new ways to use nature.*

However, we face another problem. As most new scientific researches are normally carried out in universities, while most new commercialization efforts are undertaken in industry, conflicts of interest do exist between the two parties. We have the pursuit of knowledge and new discoveries by university scientists on the one hand, and the need to generate revenue and profits for shareholders and CEOs on the other. How can they cooperate so as to simultaneously advance science and develop technology?

Here is an important fact about R&D: nations which do the most R&D have the largest economies in the world. National Gross Domestic Product (GDP) correlates with national R&D, as demonstrated in Figure 1.1, which shows the levels of global research & development (R&D) in areas with high GDP figures. One can see that in North America ($492 billion), Europe ($367 billion), and East and Southeast Asia ($614 billion), these countries have significant R&D expenditure, in the range of hundreds of billions of U.S. dollars.

The next most interesting thing about R&D is that it is performed on a national scale. Plotting the size of a nation's R&D against its GDP shows vividly that unless a nation invests in R&D, its economy cannot effectively compete on a global scale (see Figure 1.2).

After the Second World War, the United States has become the largest investor in R&D (science & technology) and has continued doing so into the 21st century, with an R&D expenditure of over $450 billion dollars. The European Union nations also have some of the highest GDPs in the world, and an R&D expenditure of $340 billion. Nevertheless, 40 years of an 'open-up' policy and economic reforms instigated by Deng in the 70s,

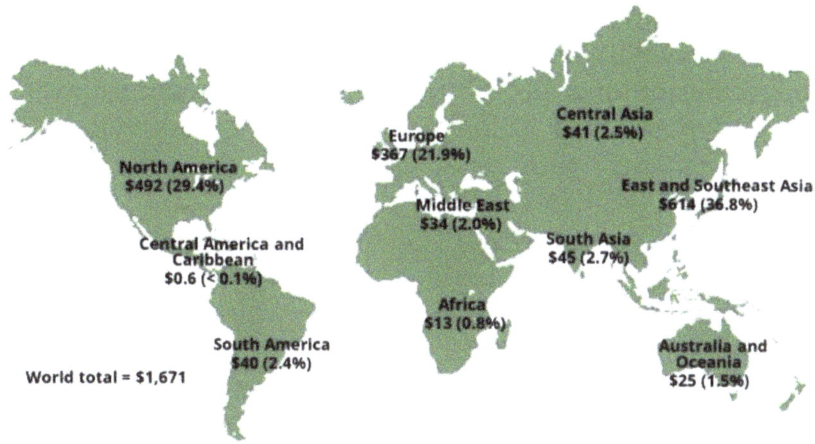

Figure 1.1. Global R&D expenditures, by region in 2013.

Source: NSF Science Indicators 2016.

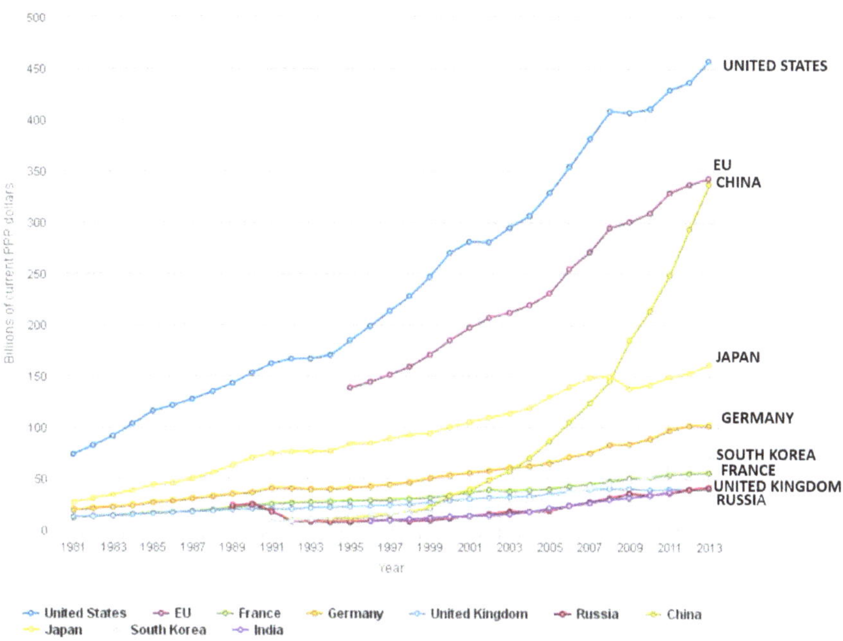

Figure 1.2. Expenditures on R&D 1981–2013.

Source: NSF Science Indicators 2016. https://www.nsf.gov/statistics/2016/nsb20161/#/

have propelled China to become the world's second largest economy (by nominal GDP) and the world's largest economy (by purchasing power parity) according to the International Monetary Fund (IMF). China's state companies also invested in R&D, so much so that by 2013, China's R&D expenditures equaled those of the EU.

Expenditure on R&D is a significant cost to a nation. For a nation to be globally competitive, its R&D expenditure needs to rest in the range of 2–4% of the nation's GDP (R&D/GDP) (see Figure 1.3).

Over the last two decades, the United States' expenditure ratio (R&D/GDP) has been around 2.5%, while Germany's ratio climbed

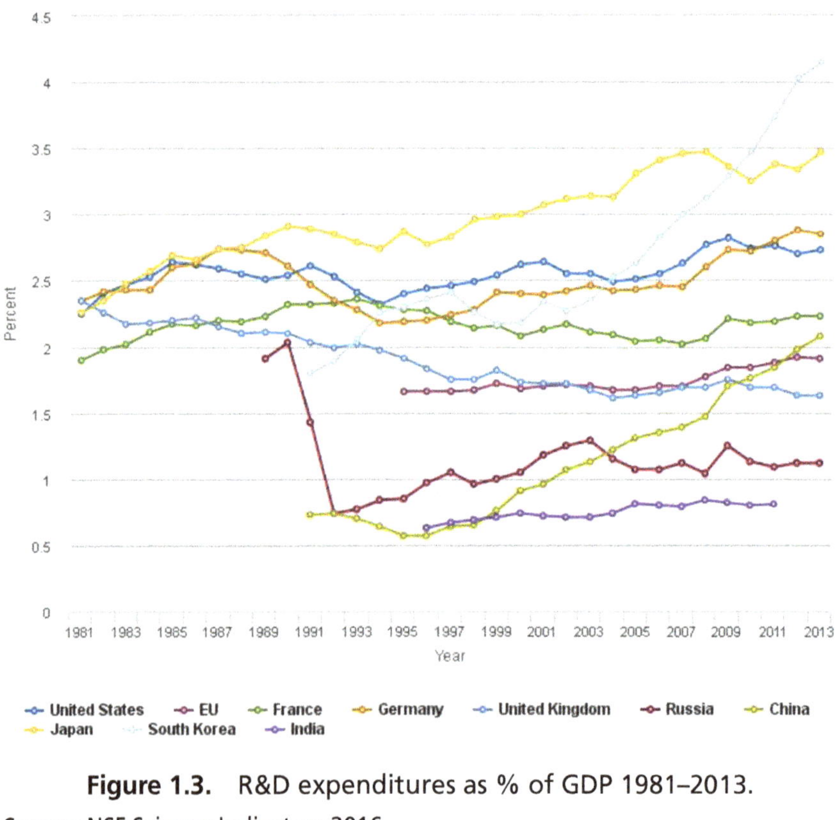

Figure 1.3. R&D expenditures as % of GDP 1981–2013.

Source: NSF Science Indicators 2016.

to 3.5%, South Korea's to over 4%, and Japan's and China's to 2%. On the other hand, the United Kingdom's ratio fell to 1.6%. India's ratio is around 0.8%, and Russia's ratio is just above 1%.

What is most striking is that the Asian countries, such as Japan and South Korea, are committing a high percentage of their GDP to R&D, since they know that science and technology are vital factors in improving their nation's global competitiveness. China, the third Asian country, is rapidly increasing its investment in science and technology. The US, Germany, France, and the United Kingdom recently are holding steady in their investment in R&D (as a percentage of their GDP). Russia and India are not seriously committed to the costs of becoming a high-tech economy.

Since R&D results in the commercialization of products, one can also see a correlation between the levels of R&D expenditure and the output of high-tech products by the different countries of the world (see Figure 1.4).

With its large R&D expenditures, the United States leads the world in the manufacturing of high-tech products at 29% output. The EU, with its large R&D expenditure, produces about 23% of high-tech manufactured products. And with the rapidly

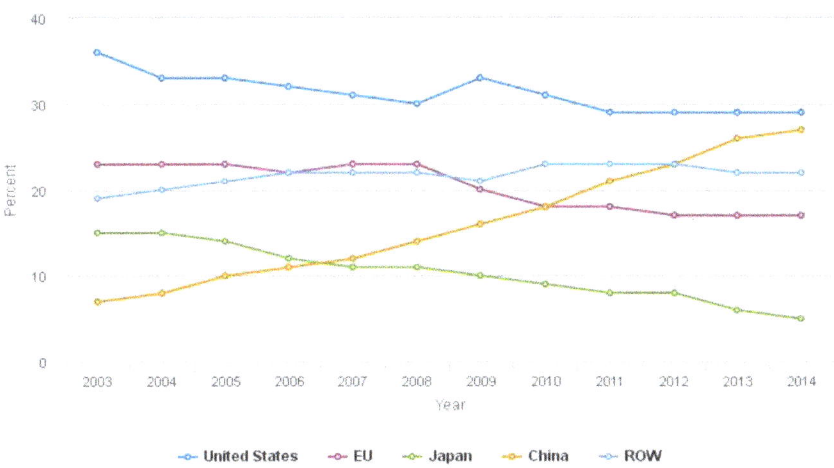

Figure 1.4. Percentage output of global high-tech industries 2003–2014.
Source: NSF Science Indicators 2016.

growing R&D expenditure in China, the Chinese share of high-tech manufactured products has grown to about 27%. (A historical footnote: the rapid growth of high-tech manufacturing in China was due to United States' industry outsourcing much of its manufacturing to China from the late 1980s, in order to reduce the cost in manufacturing, as technologies in manufacturing processes were being automated and controlled with computational technology.)

> *In summary, innovation has played and continues to play a significant role in the world's industrialization and a nation's global competitiveness. Innovation may be expensive, but it is an important investment as R&D contributes to economic growth.*

Science and Technology Policy Definitions

Let us next briefly review some of the key terms in R&D, such as *nature, science, technology, commercialization, products* — these are all central concepts to policies about innovation and economic growth. In the scientific and engineering communities, the term 'nature' is commonly used to refer to essential qualities of things that can be observed in all the universe. *(1) 'nature' is the totality of the essential qualities of the observable phenomena of the universe.*

The derivation of the term 'science' comes from the Latin term '*scientia*', meaning 'knowledge'. However, the modern concept of scientific research has come to indicate a specific approach toward knowledge, which results in the discovery and explanations of nature. *(2) 'science' is the discovery and explanation of nature. For science and technology (S&T) policies, the important institutional aspect is that science is mostly performed in universities (and not in industries or in government research labs).*

The technical side of the idea of technological innovation — invention — derives from the idea of technology. The historical derivation of the term 'technical' comes from the Greek word '*technikos*', meaning 'of art, skillful, [and] practical'. The suffix of the word 'ology' refers to the 'knowledge of' or 'systematic treatment of' '*technikos*'. Thus, the term 'technology' literally translates to 'knowledge of the skillful and practical'. This meaning of 'technology' is a common definition of the term — but too vague for expressing precisely the interactions between science, technology and economy. The 'knowledge of the skillful and practical' is knowledge of how to 'manipulate' the natural world. Technology is useful knowledge, i.e. knowledge of a functional capability. In all technologies, there is always some form of nature manipulation. Accordingly, we will use a more precise definition: *(3) 'Technology' is the knowledge of the manipulation of nature for human purposes. For S&T policies, the important institutional aspect is that technology is mostly developed in industry (and not in universities or in government research labs).*

Previously, not all technology in the world was invented upon a base of scientific knowledge. Before the emergence of science in the 1600s, previous technologies such as fire, stone weapons, agriculture, boats, writing, bronze, iron and guns were all invented before science. As a result, people with technical knowledge who understood *how* to apply, put technologies to work and made inventions. To improve technologies, one must understand how they functioned. **What science does for technology is explain why technologies work**. After the conception of science, all important technologies in the world have been invented upon knowledge based on science. *(4) 'Scientific technology' is a technology invented upon a scientific base of knowledge, which explains why the technology works. For S&T policies,* ***the important institutional aspect is that scientific technology is most rapidly advanced through university-industry research cooperation.***

Technologies are implemented in products and services by designing and embedding the technical knowledge into the operation of the product/service. Engineers do the designing; their designs enable businesses to use nature (through technology) to add value to products. *(5) 'Engineering' is the design of economic artifacts embodying technology. For S&T policies, the important institutional aspect is that the* ***engineering design of products is mostly performed in industry****.*

Products and services provide utility to the customers who purchase them. Through products/services, the concept of 'utility' provides the functionality of a technology to a human for a purpose. Thus, economic utility is created by a product or service sold in a market that provides a functional value for its customer. Take for example xerography products, which provide the function of copying (duplicating) the contents of printed papers to the customer, at a much quicker rate than copying by hand.

In a society, its technology connects nature to its economy; hence, we will define the term 'economy' to indicate this. The common usage of the term 'economy' indicates the administration, management or operation of the resource(s) and productivity of a household, business, nation, or society. But we will use the term 'economy' in the following way: *(6) 'Economy' is the social process of the human use of nature as utility. For S&T policies,* ***the important institutional aspect is that economic growth is mostly achieved by industry (and not by universities or the government).***

Business organizations provide the social forms for economic activities. The leadership in an economic organization is provided by the management staff of the business. *(7) 'Management' is the form of leadership in economic processes. For S&T policies,* ***the important institutional aspect of industrial management is that it serves both private and public good in a national economy****.*

In modern economies, the direct connection between technology and economy are in the products, processes and services that embody change in the technologies, which a business uses in an economy. (8) *'high-tech products/services/processes'* are commercial artifacts which operate on the principles of a new technology.

Civilization of Scientific Technology

The definitions of science and technology are important because they are factors at the heart of our present civilization. To understand modern civilization, one needs to see the significance of innovation in economic development. Modern innovation occurs within 'national systems of innovation', which perform significant R&D.

After the inception of science in the 1600s, all new major technologies after 1700 have been invented on the basis of science knowledge. Scientific progress is essential to technological progress (and vice versa). Figure 1.5 lists some historically important innovations. One can see that most important innovations occurred after 1700.

This dramatic increase in the rate of technological innovation was due to the emergence of science in Europe in the 1600s. With science, inventing new technologies was made easier and more attainable. This phenomenon, which we now call 'scientific technology' is technological innovation made upon a scientific base of knowledge. Note how rare technological innovations were up until the 1700s — e.g. the discovery and invention of tools, pottery, bronze, writing, iron, guns, printing, and telescopes. Then, note the number of technological innovations after the 1700s as a result of science (see Figure 1.5).

In addition, Figure 1.6 shows the timeline of technological innovations, underpinned by the progress in science and technology. Science began in Europe in the 17th century when Newton combined new ideas of physics (from Copernicus, Brahe, Kepler, and Galileo) with new ideas in mathematics (from Descartes and others), inventing the Newton Paradigm of Physics (which combined the mathematical theories of space, time and forces). In the next century, these ideas were further expanded into new scientific disciplines, namely physics, chemistry, and mathematics. The 1800s and 1900s found dramatic advances in these disciplines, along with the founding of the scientific discipline of biology. At the end of the 1900s, the physics of the small

HISTORICALLY IMPORTANT INNOVATIONS

INNOVATION	FUNCTION	DATE
TOOLS	TECHNOLOGY	PRE-HISTORY
POTTERY	MATERIALS	STONE AGE
BRONZE	MATERIALS	2500s BC
WRITING	LITERACY	2500s BC
IRON	MATERIALS	1500s BC
GUN	WEAPONS	1300s AD
PRINTING	LTERACY	1400s
TELESCOPE	OPTICS	1500s
MICROSCOPE	SCIENCE	1700s
STEAM ENGINE	POWER	1700s
POWERED MACHINERY	PRODUCTION	1700s
RAILROADS	TRANSPORTAION	1830s
TELEGRAPH	COMMUNICATIONS	1850s
CHEMICALS	MATERIALS	1850s
STEAM SHIPS	TRANSPORTATION	1860s
CAMERAS	IMAGES	1860s
TELEPHONE	COMMUNICATIONS	1880s
ELECTIRC LIGHTING	ILLUMINATION	1880s
ELECTRICAL POWER	POWER	1880s
BICYCLES	TRANSPORTATION	1880s
AUTOMOBILES	TRANSPORTATION	1890s
AIRPLANES	TRANSPORTATION	1900s
PLASTICS	MATERIALS	1900s
MOVIES	COMMUNICATIONS	1910s
ELECTRON TUBES	ELECTRONICS	1910s
RADIO	COMMUNICATIONS	1920s
RADAR	SENSING	1930s
SPACE ROCKETS	TRANSPORTATION	1930s
NUCLEAR FISSION	WEAPONS	1930s
TELEVISION	COMMUNICATIONS	1930s
COMPUTERS	COMPUTATION	1940s
TRANSISTORS	ELECTRONICS	1940s
SATELLITES	TRANSPORTATION	1950s
INTEGRATED CIRCUITS	ELECTRONICS	1950s
COMPUTER NETWORKS	COMMUNICATIONS	1970s

Figure 1.5. Historically important innovations.

parts and largest spaces of matter were established, along with the chemistry of inanimate and animate matter, the molecular biology to explain the inheritance of life, to the extension of computational science of mind and communication. All these developments began internationally from the very first demonstration of science, and hence the last four centuries can be labeled as a period of the *globalization of science*.

In contrast to the international growth of science, the economic and technological developments occurred within purely

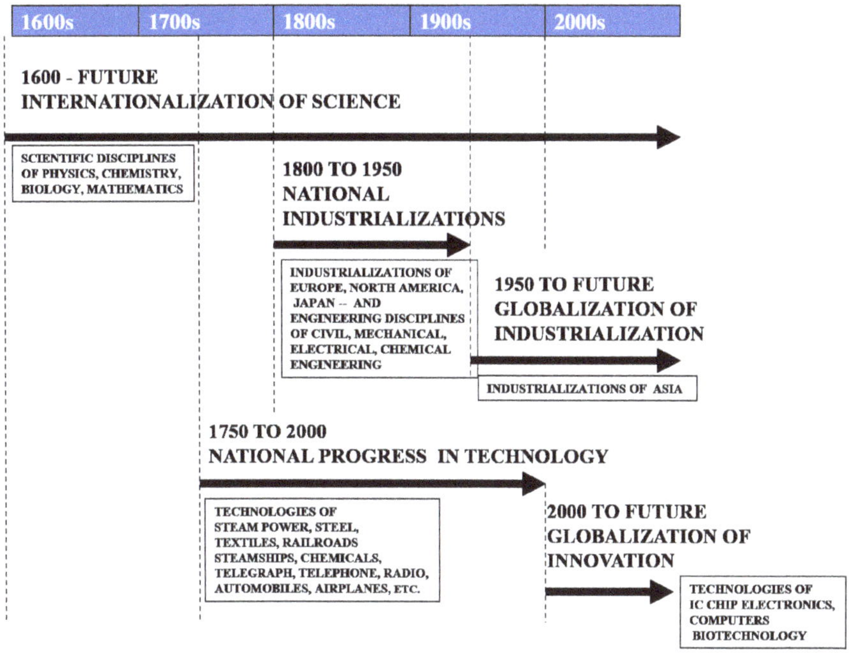

Figure 1.6. Time-lines of science, technology, and economy.

national contexts. Each nation industrialized on a national basis and in competition with other nations. From about 1765 to about 1865, industrialization occurred mainly in the European nations of England, France, and Germany. For the next one hundred years, from 1865 to about 1965, other European nations began to industrialize; but the principal industrialization had by then shifted to North America.

By the middle of the 20th century, in the 1940s, the US's industrial capacity alone was so large and innovative, to the point that it became a determining factor in the conclusion of the Second World War. For the second half of the 20th century, USA's industrial prowess continued, and European nations rebuilt their industries that had been destroyed by the War. From 1950 to the end of the last century, several Asian countries began to emerge as globally competitive industrial nations: Japan, Taiwan, South Korea and Singapore.

After the economic reforms in China by Deng Xiaoping, China began to industrialize rapidly, and quickly became a major manufacturing nation in the 21st century. This was assisted by global firms in the United States, which had begun outsourcing manufacturing to China, particularly in consumer electronics. The communist government in China absorbed the cost of building new manufacturing facilities in China, to the standards of American designs.

India, after beginning to throw off decades of socialism, also began to industrialize, particularly in information technologies. Many other Asian countries were also moving toward globally-competitive capability: South Korea, Vietnam, Thailand, Philippines, Malaysia and Indonesia. (Note that, historically, Asian industrialization actually begun in Japan in 1865, but was diverted to a military-driven society and industry. After World War II, a re-industrialization of Japan occurred.)

In summary, over the last three hundred years of world industrialization, different regions of the world began to develop globally-competitive industries over three waves:

(1) The first hundred years (1765–1865) — Europe
(2) The next hundred years (1865–1965) — North America
(3) The latest hundred years (1965–2065) — Asia

With technology viewed as a national asset, technological progress on a national basis developed with the industries. The first technological inventions, which kicked off the Industrial Revolution, took place in England, with the invention of the steam engine, coke-fueled steel production, textile machinery and factories. These provided the basis for the Industrial Revolution. Many new technological innovations followed — chemicals, materials, communications, transportation, electronics, computers, etc. — all occurring in national contexts. In the 1900s, innovation in new technology became the basis of national economic competition.

However, the rapid international transfer of modern technology accelerated changes in the second half of the 1900s. The beginning of the 21st century resulted in a new pattern of change in the modern world, resulting in the emergence of the *globalization of technological innovation.*

Thus, by the end of the 20th century, there was widespread global appreciation for science and technology, which are critical to international economic competitiveness. World markets and industrial production had become global affairs. Global trade in 1980 had already accounted for about 17% of total economic activity, increasing to 26% worldwide by 2000. The economic mechanism of global trade was multi-national firms (MNCs): "Global trade increased rapidly throughout the 1990s, as multinational companies shipped products through a global supply chain that minimized costs and maximized efficiency with little regard for national borders." (Kahn, 1990).

As the entire world was industrializing, it was important to clearly distinguish between globally effective and ineffective industrialization. For example, Michael Porter (1990) noted several factors in effective national competitive structures, including: political forms, national and industrial infrastructures, domestic markets, and firm strategies. An effective national research infrastructure was also necessary for effective industrialization. Elements of necessary national infrastructure include: educational systems, police and judicial systems, public health and medical systems, energy systems, transportation and communication systems. Science and technology policy and the economic development of a nation in this global context remain important problems. Technological progress has enabled some but not yet all nations to develop economically.

An important research feature for national competitiveness lies in the development of the right strategic interactions between universities and high-tech companies in the nation. For example, Peter Gwynne (1996) described some of the science and technology parks developed in Singapore, South Korea, and Taiwan, that bring their science and technology infrastructure for high-tech industries closer together. These science and technology parks were modeled after the history and geographical region of Silicon Valley in northern California in the US (devoted to the building of the personal computer and computer chip industries). Stanford University and the University of California at Berkeley both played an important role (along with venture capital firms) in growing the high-tech industries of chips, computers, and multimedia.

Summary

The innovation challenge for the 21st century is how to effectively encourage university and industrial research cooperation on a global scale — within global corporations (MNCs) and in small-medium enterprises (SMEs) — but with universities in competing national contexts.

The trend in cooperative innovation has been this:

- Universities are the primary source of scientific progress.
- Industries are the primary source of technological progress.
- Governments are the primary source of funding for science and the creation of conducive environments to promote the development of radical innovations.
- Science and technology interact to drive progress in both science and technology.
- Research cooperation is important as no nation can be economically successful in the global market without being competitive in both science and technology.

Interacting together, progress in science and technological is derived from discovery and invention. Industry transforms invention into utility using R&D activities based on scientific knowledge.

In the global world, we will examine how different research sectors could cooperate and foster national strength and competitiveness using Science, Technology and Innovation.

CHAPTER 2

National Innovation Structure

Introduction

As emphasised in Chapter 1, technological innovation consists of science, invention, and the commercialization of new technology. In this chapter, we should first examine the institutional structure for innovation in a nation. At the national level, this is conducted by a Science and Technology infrastructure called a '*national innovation system*'. At the macro level, this involves sharing knowledge across research sectors in a country. At the industrial level, this involves the sharing of knowledge across the firms in an industry. At the micro level, individual firms innovate by researching and designing new technologies in new products, services, production, and operations. The international competitiveness of a country is decided by the interaction between the macro and micro levels of innovation in the country.

Richard R. Nelson (1993) wrote:

> "Today, R&D facilities, staffed by university trained scientists and engineers attached to business firms, universities, or government agencies, are the principal vehicles through which technological advance proceeds... An important characteristic of technical advance... is that it is strongly supported by various fields of science... By the beginning of World War I, the industrial research laboratory, a facility dedicated to research and the development of new or improved products and processes, and staffed by university trained scientists and engineers, had become the principal locus of technical advance in the chemical and electrical industries, and was beginning to become

> important more and more widely. The industrial laboratories were teamed with universities that trained their new R&D scientists and engineers... [who] undertook research in the new applied sciences and engineering disciplines as well as in the more traditional basic sciences."

The first issue to address about innovation in S&T infrastructures is the question of when and how industry needs scientific progress. Industries do not directly utilise science, whereas technology is directly applicable to industry as it is embedded in the design of products, services, and production. However, as technologies mature, industries need new science to continue making progress in technology. This progress cannot be made without a deeper understanding of the science unpinning the technology. Science is also needed to create the scientific basis for new technology. To enable innovation, industries and universities need to collaborate in research.

Radical vs. Incremental Innovation

In the early studies on innovation back in the 1960s, Don Marquis (1969) identified three types of innovation: *incremental, radical, and systems*. Today, it is more useful to divide innovations into these three types: *incremental innovation, radical component-innovation and next-generation technology (NGT) systems*. This strategy enables universities to participate more in radical and next-generation technology innovations, rather than in incremental innovations. An *incremental innovation* is a small change in the performance of an existing technology; while radical innovation creates a new technology with radical changes in performance and function. For incremental progress, industries can usually do the needed basic research as applied science in industrial laboratories. However, there are many more incremental innovations than there are radical innovations, components or systems, and both are vital to our economic progress. Most radical innovations begin in universities. The distinction between incremental and radical innovation is important to our understanding of how government research, university research and industrial research could be effectively integrated together.

Case Study: The Radical Innovation of the Internet

The innovation of the Internet provides an excellent example of radical system innovation, and we will review the early history of Internet innovation which started in the 1970-80s. As an innovation, the Internet is both an *idea* of a technology and an *implementation* of the technology as a connected set of businesses. At any instant, the architecture of the Internet is constructed out of many, many units which connect to or disconnect from the network at different times — either as businesses directly connecting to the Internet or as home-based users connecting to surf the web. Internet technologies, as a functional system, enable users (whether businesses or residential users) to communicate with other computer owners by logging in to the internet through personal computers or web servers.

From the industrial perspective, the Internet was commercialised by a group of businesses which supply the items and services a person would need to connect to the Internet, as well as the services they can enjoy once they are connected. Some examples of these items and services are:

1. A personal computer (e.g. Dell), containing a microprocessor (e.g. Intel CPU), an operating system (e.g. Microsoft Windows), and a modem/wifi,
2. An Internet service provider (e.g. AOL, Vodaphone, Comcast, etc.)
3. A server and router (e.g. Cisco, Dell, IBM, etc.)
4. A local-area network or wide-area network in a business (e.g. Cisco, Erickson, etc.)
5. An Internet back-bone communications system (e.g. AT&T, Sprint, Vodaphone, etc.)
6. Internet search engines (e.g. Google, Yahoo, etc.)

7. Businesses services (e.g. Amazon, eBay)
8. News and entertainment services (e.g. Nytimes.com, Netflix)
9. Social networks (e.g. Facebook, Twitter, etc.)
10. Email services (e.g. Yahoo, Google, government networks, etc.)

The arrival of the Internet can be traced back to an earlier computer network called the ARPAnet, which could in turn be traced to Dr. J.C.R. Licklider, who served in the Advanced Research Projects Agency, working on military research projects funded by the U.S. Department of Defense. In ARPA, he spearheaded research on how computers could be used for military command and control (Hauben, 1993). Licklider began developing a way to connect computers together in a network that enabled them to communicate with each other by writing a series of memos on his thoughts about networking computers and sharing them with the computer science research community. At the same time, university researcher Leonard Kleinrock came up with another key idea in computer networking: sending information in packaged groups, or packet switching. He published the first papers on packet switching in 1962 and 1964. Packet switching paved the way for sending messages swiftly in bursts of data without tying up communication lines between the two parties during the entire period of transmission. This innovation marked the move away from close-coupled computer networks to loosely-coupled computer networks.

In 1965, Lawrence Roberts from the Massachusetts Institute of Technology (MIT) connected a computer at MIT to one in California through a telephone line. In 1966, Roberts submitted a proposal to ARPA to develop a computer network for military defence purposes; to protect U.S. Military communications in case of nuclear attacks. This was called the Advanced Research Projects Administration Network, or the ARPAnet. Robert W. Taylor replaced Licklider as program officer of ARPA's Information Processing Techniques Office. Taylor had read Licklider's memos and was also thinking about the importance of computer networks. He funded Roberts's ARPAnet project: "The Internet has many fathers, but few deserve the label more than Robert W. Taylor. In 1966... Taylor [funded the] idea for Internet's precursor, the ARPAnet." (Markoff, 1999)

Earlier, Taylor had been a systems engineer at the Martin Company, working as a research manager at the National Aeronautics and Space Administration (NASA), funding advances in computer knowledge. Then, he went to ARPA and became interested in the possibility of communications between computers. In his office, there were three time-sharing computers in three different research programs that ARPA was supporting. He watched communities of people build up around each time-sharing computer: "As these three time-sharing projects came alive, they collected users around their respective campuses... [But]... the only users... had to be local users because there was no network... The thing that really struck me about this evolution was how these three systems caused communities to get built. People who didn't know one another previously would now find themselves using the same system." (Markoff, 1999)

Taylor was also struck by the fact that each time-sharing computer system had its own commands: "There was one other trigger that turned me to the ARPAnet. For each of these three terminals, I had three different sets of user commands... I said... it's obvious what to do: If you have these three terminals, there ought to be one terminal that goes anywhere you want to go where you have interactive computing. That idea is the ARPAnet." (Markoff, 2000) In 1965, Taylor proposed to the head of ARPA, Charlie Herzfeld, the idea for a communications computer network, using standard protocols. Thereafter, in 1967, a meeting was held by ARPA to discuss and come up with a consensus on the technical specifications for a standard protocol for sending messages between computers. These computers were called 'Interface Messaging Processors (IMPs)" and were used to design messaging software. The first node on the new ARPAnet was installed on a computer on the campus of the University of California at Los Angeles. The second node was installed at the Stanford Research Institute, and the ARPAnet began to grow from one computer research setting to another. By 1969, the ARPAnet was up and running. Taylor left ARPA to work at Xerox's Palo Alto Research Centre.

As the ARPAnet grew, so did the need to control the system. It was decided that it would be controlled by another protocol, called the Network Control Protocol (NCP); this decision was made in December 1970 by a private committee of researchers called the Network Working Group. Thus the ARPAnet grew into an overall interconnected, but independent, set of multiple smaller networks. In 1972, a new program officer at ARPA, Robert Kahn, proposed an advance in the protocols for communication. The new protocols would be 'open-architecture', or accessible to anyone. The Transmission Control Protocol/ Internet Protocol (TCP/IP) was invented, and became the open standard upon which the world's Internet would later be based.

While the ARPAnet expanded in the 1970s, other government agencies and universities were also hard at work in developing technology for computer networks. In 1981, the National Science Foundation (NSF) established a supercomputer centres program, which provided researchers throughout the U.S. with access to their supercomputers at the five NSF-funded supercomputer centres. It also provided support to the same researchers. In 1989, a researcher, Tim Berners-Lee, who was working at Conseil Européen pour la Recherche Nucléaire (CERN) at the time, came up with the idea of sending structuring information over the Internet. He used a HTML script with Uniform Resource Locators (URLs) to link one website to another as the WWW over the Internet. Berners-Lee then wrote the first web browser computer program in 1990.

Earlier in 1988, a committee of the U. S. National Research Council had been formed to explore the idea of an open, commercialized Internet. They sponsored a series of public conferences at Harvard's Kennedy School of Government on the "Commercialization and Privatisation of the Internet".

In April 1995, NSF stopped supporting its own NSFnet 'backbone' of leased communication lines. From then on, the Internet was privatised. The Internet grew rapidly to connect over 50,000 networks all over the world. On October 24, 1995, the Federal Network Council defined the Internet as:

— Logically linked together by a globally unique address space based on the Internet Protocol (IP),
— Able to support communications using the Transmission Control Protocol/Internet Protocol (TCP/IP) standards.

The Internet was an innovation created from government-university-industry research cooperation.

Innovation and R&D Infrastructure

Venn diagrams can be used to model an S&T national innovation system (see Figure 2.1). The function of a national innovation system is to use knowledge from research to create technological utility, as shown by the vertical arrow connecting the Venn diagram of the macro-level of innovation to that of the micro-level of innovation.

Innovation is the process of conducting scientific/technological research (knowledge) and using the results to create products/processes/services of national and economic usefulness (utility). An institutional infrastructure for innovation must include at least three kinds of organizations:

(1) Universities for scientific advances and technological invention,
(2) Industry for technological invention and commercialization, and
(3) Government research to provide a platform and framework for the advancement of science and military technology. Science is costly but it is for the public good, hence the government funding of scientific research in universities is

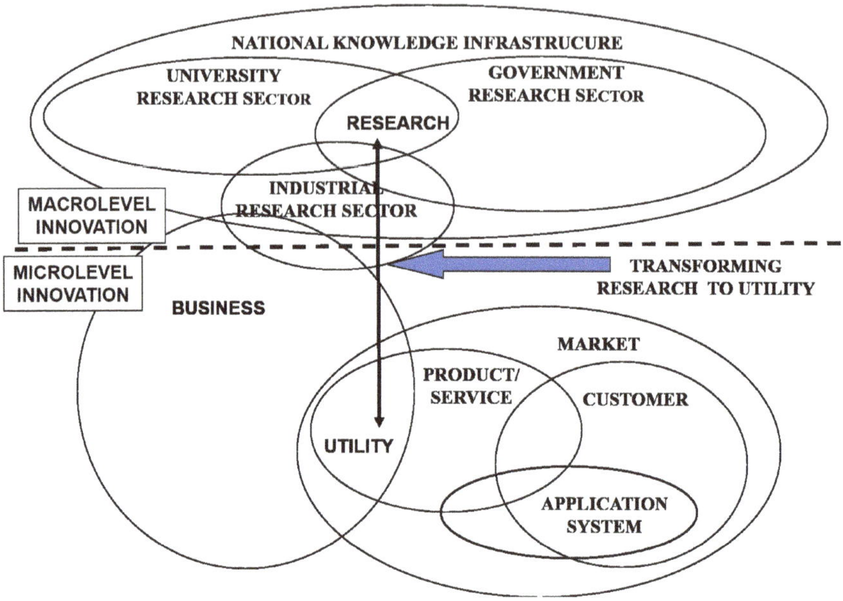

Figure 2.1. The national innovation system.

vital and essential. New businesses also sometimes emerge from university research, spinning off directly into new-venture start-up companies.

- *The research sectors of universities and government provide a macro level of innovation policy in a nation.*
- *The industrial sectors of high-tech businesses provide a micro level of innovation strategies.*

Nelson (1993) also wrote:

> "Universities in most countries are, first, the places where much of the basic research in fundamental sciences such as physics is undertaken, although the reliance on universities as a locus of basic research, as contrasted with national laboratories, varies across countries. Research in basic sciences such as physics tends to be guided by the internal logic of the discipline rather

than by expectation of particular practical application... In almost all nations, universities now are funded to a substantial degree by governments. However, the organization and means of funding, and orientation of university research, differ significantly across nations... Government laboratories also are an important part of many national innovation systems. In the United States, government laboratories play important roles in... the fields of agriculture, health, and nuclear energy... In other countries public laboratories play a significantly larger, or different, role... the Federal Republic of Germany has a network of Max Planck Institutes dedicated to basic research, and Fraunhofer Laboratories dedicated to applied research. Much of the latter's work is aimed at helping industry."

In Figure 2.2, we can see the example of the Internet and its development, depicted in the national innovation system.

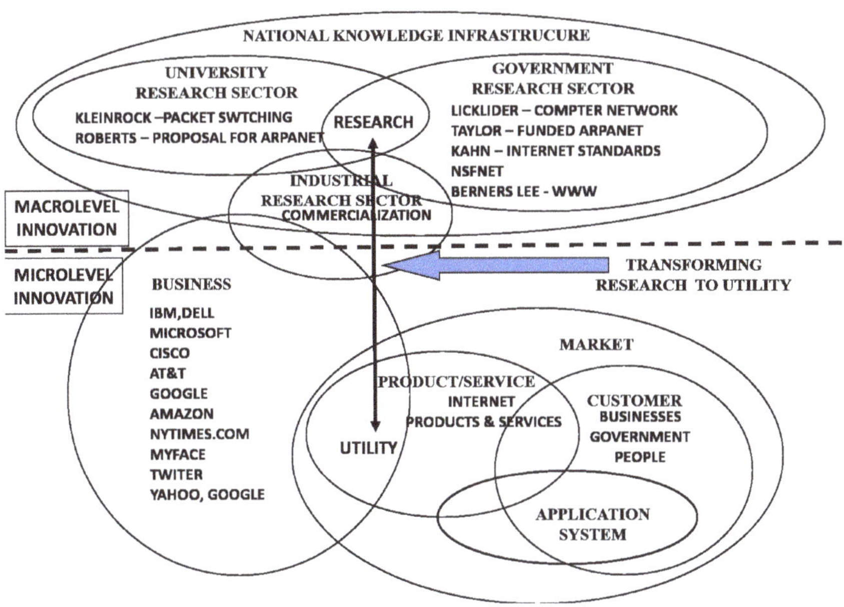

Figure 2.2. Internet innovation & development in the U.S. national innovation system.

Figure 2.2 emphasizes contributions from government and university research to the development of the Internet — followed by industrial research and the commercialization of the Internet through product and service innovation offered by Internet businesses.

The innovation of the Internet occurred at the macro level, motivated by researchers seeking ways to have computers communicate with each other, and created a new kind of functional capability in computation. The invention of computer networks required the creation of several technical ideas, and together these ideas laid the foundation for the development of technology that led to the creation of the Internet. Some of these ideas are:

1. *Computer-to-Computer Communications*, where computers would be electronically connected to each other.
2. *Packet-Switching*, where computer messages should be transmitted in brief, small bursts of electronic digital signals — rather than a continuous connection used in the preceding human voice telephone system.
3. *Standards*. Formatting the digital messages between computers needed to standardize for the sending and receiving of message in the forms of packets, and these open standards became the Internet's (TCP/IP) standards.
4. *Routing*. A universal 'address' repository would provide unique addresses, in the form of IP addresses, so that computers could know where to send messages to one another.
5. *HTML*. Web pages would be written in a language which allowed computers to link to other sites
6. *www. Web Page Publication*. Software facilitated the preparation and publication of sites on the Internet. Registration of directory of web sites allows sites to be connected through the Internet.
7. *Browser*. Software on computers allowed users to link to the World Wide Web (www) and find sites.

8. *Search Engine*. Software allowed users to search for relevant sites and link to them.
9. *Social Media*. Software enables users to form virtual social groups for the sharing of contents.

Together, all these technical ideas enabled the development of the new Internet technology. The commercialisation of technology occurred when the NSF transferred network management from the government to private companies. The innovation of the Internet occurred in a pattern of technological innovation:

1. First the *science* of computers (computer science),
2. Next the *invention* of new technical ideas (such as the ARPAnet) and
3. Finally the *commercialization* of new products and services embodying these new ideas (in the privatization of the Internet).

The process outlined above demonstrates that the pattern of innovation is ***science-invention-commercialization.***

Quadruple Helix — National Innovation Systems Over Time

In 2000, H. Letzkowitz and L. Leydesdorff introduced a time dimension of interactions within this national system infrastructure, calling such interactions a kind of 'triple helix of university-industry-government relations' (Letzkowitz & Leydesdorff, 2000).

In 2009 and 2010, E.G. Carayannis and D. F. J. Campbell extended this triple helix model to a quadruple-helix by adding another interaction to the innovation process. This fourth interaction stems from the civic public, and takes the forms of media and culture (Carayannis & Campbell, 2009) (Carayannis &

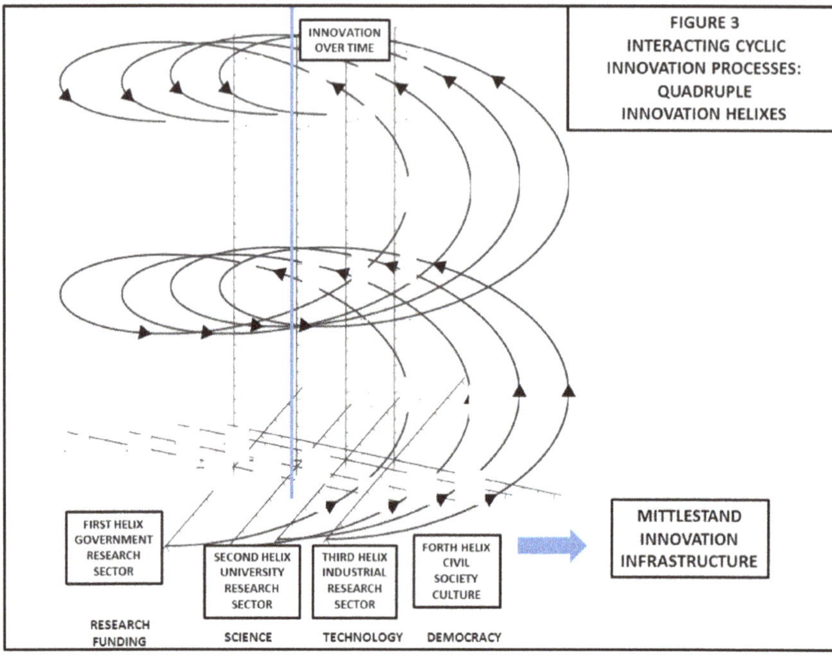

Figure 2.3. The quadruple helix of the temporal innovation interactions.

Campbell, 2010). In 2012, they expanded upon the elements of the fourth helix of civil society and culture, as shown in Figure 2.3 (Carayannis & Campbell 2012). The addition of the fourth helix indicates that the culture of a nation can impact the three helical interactions between government, industry, and universities.

The use of the four helical forms to depict the processes of innovation emphasises that these institutional processes occur in the three research sectors of government, industry and university, facilitated by a civil society and culture. It also emphasizes that the institutional processes mentioned are all cyclic in two ways. The knowledge processes in each sector are cyclic, and so is the passing of knowledge between sectors, each respectively focused upon 'funding,' 'science', 'technology', and

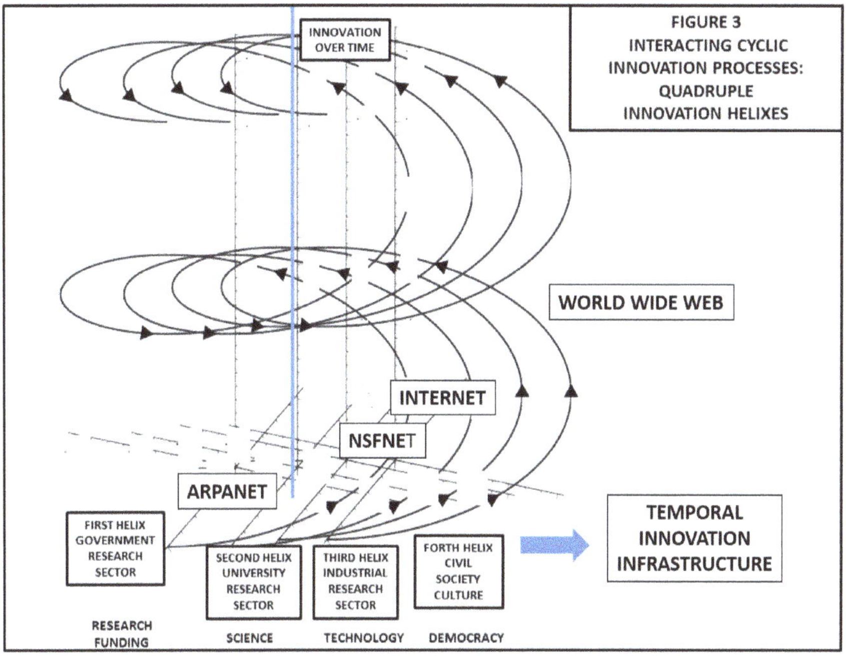

Figure 2.4. The quadruple helix of the WWW internet innovation.

'culture.' This model emphasises that institutional interactions in a national innovation system are not linear, but circular.

Figure 2.4 shows how, over time, the helix drove the innovation of the World Wide Web on the Internet.

Here are the cyclic interactions that occurred in the US national innovation system and led to the innovation of the Internet:

— First came the government-university contribution of ARPANET, then
— The government-university contribution of the NSFNET, and
— Finally, the industry commercialization of the INTERNET. Societal and business contributions filled out the INTERNET with the WORLD WIDE WEB and VIRTUAL NETWORKING.

U.S. National Innovation Structure

Total R&D expenditures around the world were around $1.671 trillion in 2013, rising from $1.269 trillion in 2008 and $836 billion in 2003. Fifteen countries/economies expended $19 billion or more on R&D in 2013, accounting for 86% of the global total; the top rankings were dominated by the United States and China.

- North America expended 29.4% of R&D,
- East and Southeast Asia 36.8%,
- Europe 21.9%.
- The expenditures of other areas were around 2% or less.

In the United States, R&D expenditures had risen from about $40 billion dollars in 1953 to $440 billion dollars in 2013. In the U.S., 71% of R&D expenditures were by industry, 14.2% by universities, 10.9% by government, and 4.1% by non-profit organizations. According to data on the sources of U.S. funding of R&D, industries provided 65% and the government 24%. (NSF Science Indicators 2016)

The breakdown of how the U.S. government's R&D expenditures are invested, by percentage, is as follows:

- 45% into basic research,
- 41% in applied research, and
- 14% in development.

The R&D performed domestically by U.S. businesses occurred mainly in five industrial sectors:

- Chemical manufacturing (particularly the pharmaceuticals industry);
- Computer and electronic products manufacturing;
- Transportation equipment manufacturing (particularly the automobiles and aerospace industries);

- Information (particularly the software publishing industry); and
- Professional, scientific, and technical services (particularly the computer systems design and scientific R&D service industries).

In 2013, these five industrial sectors accounted for 82% of the $322.5 billion of total domestic business R&D performance that year.

Within the U.S., of all domestic R&D expenditure in 2013:

- 82% was mainly funded through performing companies' own funds.
- For the remaining 18%, where the R&D is performed by companies but funded by others, the largest source of funding is the federal government, whose funding accounted for about 9% of the business R&D performance total in 2013.

During the latter half of the twentieth century, the economic world progressed from a world of international trade into a world of regional markets (e.g. European Union) serviced by global corporations. While science has always been international, technology was primarily national through the twentieth century. Now in the global economy of the twenty-first century, both science and technology have become global. Innovation now occurs within an international system of innovation.

Nelson (1993) wrote:

> "There is good reason to believe that in recent years, just as the idea of national innovation systems has become widely accepted, technological communities have become transnational as never before. There has been, first, a strong trend for manufacturing business firms to become multinational. This trend was set, initially, by American based firms; somewhat later European and Japanese firms joined in. In the 1980s

> international joint ventures on particular product design and development projects, or on large scale research projects, began to crop up in a number of industries. These developments partly result from and certainly reinforce other trends toward internationalization of technology. Engineers and applied scientists now are taught pretty much the same thing in schools in different countries. The dramatic lowering of national barriers to trade following the war, and the recent convergence of living standards and factor prices in the major industrial nations, means that, increasingly, firms face roughly the same market environment wherever their home base. For both of these reasons, by the mid-1980s the technologies known and employed in the major industrial nations were pretty much the same in most industries."

Summary

Scientific enquiry and technological innovation are institutionalized within the nation, creating a conducive environment for R&D. This environment acts as a test-bed and incubator that allows universities, government, and industry to interact, explore, and invent. At the macro level of a nation, it is the interaction of these sectors that has given rise to innovation in product and service design. The example of the Internet exemplifies a radical innovation in this modern-day technological revolution. Interactions in the innovation structure occur over time. As these interactions are circular, they form a kind of helix over time.

And in that helical circularity, one sees that all three research sectors are important to innovation and national competitiveness. Since it is the universities, and not the industries, that are the principal driving force behind this scientific progress, industries must look to universities for progress in science, even if these industries do not traditionally advance science. This scientific progress must come in forms that industries can directly use, and it must adhere to a timeline that they find convenient. The practical issue in national S&T policies is the development of an effective S&T infrastructure to sharpen national competiveness. The key questions are:

(1) How can firms use universities to stay technically competitive in a world of rapidly changing science & technology?
(2) How can universities obtain funds, from both government and industries, to support advancements in science, while also being responsive to industrial needs for science?
(3) How can a government best direct its R&D effort to facilitate partnerships between industry and universities?
(4) How can knowledge be shared across nations and throughout the world to promote scientific and technological know-how, but not at the expense of nations that make breakthroughs and discoveries?

We have reviewed the S&T concept of the *structure* of a national innovation system, and next we shall examine the concept of the *process* of innovation in a nation.

Chapter 3

Innovation Process

Innovation Process

In Chapter 2, we looked at the synergies created by the three R&D sectors, namely university, industry, and government, when they work hand-in-hand to prepare the ground to incubate innovation. However, innovation only occurs if the innovation process, summarized in Figure 3.1, is properly managed. Innovation as a process consists of knowledge generation which builds upon what is found in nature in order to make it useful through efforts in science, technology and product or service design. Scientific and technological advances are made in universities, products and services that consumers want are designed by industry, and innovation thrives in conducive climates created by government policies. In other words, science discovers nature and technology manipulates it, which leads to the conceptualization of new product or service ideas that are valuable to society. Industries commercialise the technology in the design of new products, processes or services. The manipulation of nature is only useful when technology is able to turn what is found in nature into a product, sold in a market as something useful to the customer — hence the process of innovation from nature to utility is a result of these collaborative efforts.

Nature becomes valuable to society only when it is useful. For example, quantum mechanics, or the field of science that explains the nature of the atom, is useful because it forms the basis for the understanding of electron movement in circuitries, which led to the use of transistors as controllable switches in

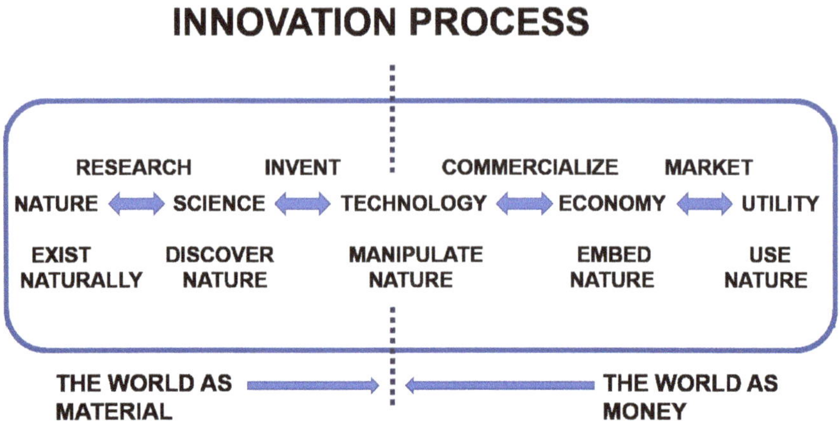

Figure 3.1. The process of innovation from nature to utility.

electronic products such as computers and cell phones, and consequently, the electronics revolution.

What is particularly striking about this process is that two different perspectives are involved:

(1) On the left side of the diagram, we see that innovation can be viewed from a technical perspective, which has to do with how innovation can be carried out.
(2) On the right side of the diagram, we see that innovation can be viewed from a financial perspective, which has to do with how profit could be made from innovation.

The innovation process requires these two perspective foci to be connected together — for technical achievements to be transformed or translated into financial returns. Entrepreneurship comes into being as the middleman to help to manage this process that connects new technology to the profits which can potentially be made.

Scientific knowledge is generated through research projects (R&D). Since science is vital to an economy, many research projects (i.e. basic research) are funded by governments, some by industry, and a few from private philanthropy. The government predominance in funding basic research is required to provide the framework for scientific research so that a way can be found to make the results of such research useful. Although some scientific research can be immediately patented, most is openly published without patents (except in scientific fields closely coupled with technological progress).

Thus, scientific knowledge of nature is used as a 'knowledge base' upon which new technologies are developed. Usually, engineers, scientists or other technical personnel invent and improve technologies. Engineers use technical knowledge to design technical function into a product/process/service/software that consumers want. Commercialization is the act of designing technology into products/services/processes and producing these solutions for sale. A business competes by selling high-tech products/services in a market place, earning income and generating profits.

As an example of how science provides a basis for technological inventions, one could look back in history to the development of molecular biology research into the biotechnology industry. In understanding the molecular functioning of the cell as the basis of life, biologists provided models of biological cells and their functioning to explain biological problems. Mark Terasaki (2016) wrote: "Students learning about the eukaryotic cell (the kind of cell of which we are composed) are taught that mitochondria are the powerhouse of the cell, the nucleus is the information storehouse, and lysosomes are the garbage disposal." The terms 'mitochondria', 'nucleus', and 'lysosomes' are names of parts of a cell. The terms 'powerhouse-of-the-cell', 'information-storehouse', and 'garbage-disposal' are descriptions of how the parts of the cell function to sustain life. Thus terms for components of nature provide scientific descriptions and explanations of nature, i.e. *Science discovers and describes nature.* The importance of description and explanation is that they provide a base of scientific knowledge upon which inventors can create solutions using technology to manipulate nature for human applications.

Moreover, more knowledge about nature is usually required to achieve the technological progress that can solve human problems. Terasaki (2016) wrote: "Summing up the role of the endoplasmic reticulum (ER) is more problematic, even for professional cell biologists. Traditionally, the ER is where membrane proteins, secreted proteins, and most lipids are synthesized. It is also the site of calcium regulation. More recent work shows that it is the site of antigen presentation by major histocompatibility complex class I molecules and the location of unfolded protein response, now thought to be central to several major diseases."

This is an example of how knowledge can provide utility (value) to patients. Improved understanding of the endoplasmic reticulum (ER) may lead to future biotechnological advances and the development of a cure for "several major diseases". Science ***might*** lead to technology that paves the way to the development of solutions to human problems in the forms of products or services marketed by industries. The key word is 'might' — there is no absolute certainty that any piece of scientific research will translate into technology and utility, now or even in the near future. Therefore, the government needs to be the principal funder of science because there is no certainty that any basic research might lead to the development of technology. With government funding, research can be utilised to create new technology and ultimately deliver the benefit back to the public — sometimes it pays off hugely with the emergence of new industries, such as DNA basic research and the emerging of the biotech industry!

Historical Case: Origin of the Biotechnology Industry

In here, we review and trace back the history of biotechnology, which was developed upon the knowledge base of 'molecular biology', to demonstrate how major advancements in science have prepared the ground and paved the way for technological progress. Biological systems are complex, and progress in the biotechnology industry has been limited by the lack of scientific understanding of the subject concerned.

1. How did scientific progress made in biology prepare the ground for the arrival of the new biotechnology industry?
2. How did the complexity of biological nature impact profitability in the new industry?

Historically, research identifying the cell as the basic unit of life and the study of its structure began in 1869. But the shape of the DNA module was only discovered in 1953. When James Watson and Francis Crick modelled it using Rosalind Franklin's X-ray picture of DNA, Watson and Crick saw that DNA had a double-helix form. This structure was in itself informative. It indicated the molecular action of DNA in the mitosis of cell reproduction. DNA is structured as a pair of coiled complementary strands. During reproduction, the two strands are uncoiled and separated, serving as templates for constructing new, identical DNA strands. This uncoiling and chemical reproduction of proteins is the basis of biological inheritance. Thus, the double-helical structure of DNA is molecularly responsible for the phenomenon of heredity.

Proteins serve as structural elements of a cell and as catalysts (enzymes) for metabolic processes in a cell. DNA provides the structural template for protein manufacture; replicating proteins through the intermediary templates of RNA. In a two-step biological process, DNA structures the synthesis of RNA; RNA then structures the synthesis of proteins. What remained unclear was how the information for protein manufacture was encoded in the DNA. In 1965, Marshall Nirenberg and Philip Neder deciphered the basic triplet coding of the DNA molecule. The amino acids composing the DNA structure are grouped into a unit of three acids to code for segments of protein construction.

Genes consist of groups of DNA which are used as templates for the production of molecules that serve different functions within the body of the organism which the DNA belongs to. Each gene is a section of the overall DNA string, or genome. After understanding the mechanism of the gene, the next step was trying to manipulate the gene. Several scientists began trying to cut and splice genes. In 1965, Paul Berg at Stanford planned to transfer DNA into *Escherichia coli* bacteria using an animal virus (Svrp lambda phage). *E. coli* bacteria can live in human intestines, and the SV40 virus, which can produce tumour cells in cultures of human cells, can be found in monkeys. Because of the dangerous nature of the SV40 virus, Berg decided not to proceed with the experiment, publishing a design for hybridizing bacteria in 1972. Berg organized a historic meeting on safety during the Conference on Biohazards in Cancer Research in California, U.S.A. on January 22–24, 1973 (Olby, 1974) that led to the government setting up safety standards later for biotechnology research.

A colleague at the University of California responded to Berg's request for some EcoRI enzyme, which cleaves DNA (and leaves the 'sticky ends' of the cut DNA). Berg gave the enzyme to one of his students, Janet Mertz, to study the enzyme's behaviour in cutting DNA. Mertz noticed that when the EcoRI enzyme cleaved an SV40 DNA circlet, the free ends of the resulting cut and linear DNA eventually re-formed by itself into a circle. Mertz asked a colleague at Stanford to look at the action of the enzyme under an electron microscope. They learned that any two DNA molecules exposed to EcoRI could be 'recombined' to form hybrid DNA molecules. Nature had arranged DNA such that cut ends ligated automatically.

At the same time, Stanley Cohen, who was a professor at the Medical Department of Stanford University learned of Janet Mertz's results, and also thought of constructing a hybrid DNA molecule from plasmids using the EcoRI enzyme. Plasmids are circular double-stranded DNA which float outside the nucleus in a cell and manufacture enzymes the cell needs for its metabolism, in contrast to the DNA found in the nucleus of the cell, which is principally used for reproduction. In November 1972, Cohen attended a biology conference in Hawaii. He was a colleague of Herbert Boyer at the University of California in San Francisco, who had given the EcoRI enzyme to Berg (and

Berg's student Mertz). At a dinner one evening, Cohen proposed to Boyer that they create a hybrid DNA molecule without the help of viruses. Another colleague at the dinner, Stanley Falfkow of the University of Washington at Seattle, offered them a plasmid, RSF1010, to use that confers resistance to antibiotics in bacteria so that they could see whether the recombined DNA worked in the new host.

After returning from the Hawaii Conference, Boyer and Cohen began to collaborate in joint experiments. By spring 1973, Cohen and Boyer had completed three splicings of plasmid DNAs. Boyer presented the results of these experiments in June 1973 at the Gordon Research Conference on Nucleic Acids in the U.S. (publication followed in the Proceedings of the National Academy of Sciences, November 1973). Cohen and Boyer also applied for a patent for the new technique.

After a hundred years of scientific research into the nature of heredity, beginning with Mendel, humanity finally began to understand and to be able to manipulate genetic material at a molecular level. A new industry was born — biotechnology. Boyer and Cohen would go on to win the Nobel Prize. Boyer would be involved in the first biotechnology company (Genentech) to go public and became a millionaire. The days for biologists to become industrial scientists had begun. Science has hence formed the basis for this new technology, which became valuable and eventually provided a return on investment.

Science Bases for Technology

As demonstrated in the birth of biotechnology, science has provided the knowledge bases for the development of scientific technology:

1. Scientists pursue research that asks very basic and universal questions about what things exist and how things work. In the case of genetic engineering, the science was guided

by these questions: *What is life? How does life reproduce itself?*

2. Major advances occur when a sufficient percentage of the puzzling object has been discovered and observed for someone to concretize the scientific concepts involved by figuring out how to put everything together in a model (as Watson and Crick modelled the DNA molecule). A scientific concept in the form of a model had then provided the focus for the minds of scientists and engineers alike to understand both the structure and dynamics of the process implied by the model.
3. Once science has created a new phenomenal knowledge base, inventions by either scientists or technologists may follow (for example, scientists invented the recombinant DNA techniques). These radical technological inventions will then pave the way for a new technology S-curve and begin a new industry.
4. When the new technology is pervasive across several industries (as genetic engineering is across medicine, agriculture, forestry, marine biology, materials, etc.), the technological revolution may fuel an economic expansion that gives rise to the arrival of a new industry. The long-waves of economic history are grounded in scientific advances that create basic new sectors in industrial technology.
5. From an economic perspective, science can be viewed as a form of societal investment in the possibility of future technologies. Since the process of scientific discovery is complicated and the time required for scientific discovery is lengthy, and since only the strongest will survive, science must be sponsored by government as a social 'overhead' cost for a technology to develop and grow. Without the environment put in place by the government to fund basic knowledge creation, technological innovation would not be possible.
6. There are general implications for industry. Corporations should be supportive of university research, which focuses

on the fundamental questions underlying core technologies of the corporation. Corporations need to perform some active basic research in the science bases of their core technologies to maintain their competitive edge, which stems from their position as a 'window-on-science' within the industry.

Historical Case: Origin of the Biotechnology Industry (Continued)

The biotechnology industry was created directly from scientific discoveries in genetics. Two of the first new biotech firms were Genentech and Cetus. In 1973, Cohen and Boyer applied for the basic patent on the recombinant DNA technique. Subsequently, Boyer founded Genentech, and Cohen joined Cetus. In the years from 1976 to 1982 in the United States, over 100 other research firms were formed to commercialize the new biotechnology. In 1980, Genentech and Cetus both went public; and Boyer and Cohen both became millionaires.

Genentech was founded not only by the scientist Herbert Boyer, but also by an entrepreneur, Robert A. Swanson. Swanson had heard of the new DNA technique and saw the potential for raising venture capital to start a genetic engineering firm. The story is that Swanson walked into Boyer's office and introduced himself. He proposed that they start a new firm. They each put up $500 as investment in 1976 to start Genentech. Early funding for Genentech was secured from venture funds and industrial sources. Lubrizol purchased 24% of Genentech in 1979. Monsanto bought up about 5%.

Founded earlier in 1971, Cetus had been started by a scientist, Donald A. Glaser, to screen pictures of microorganisms for rapid identification. Following Boyer and Cohen's invention of the gene recombination technique, Cetus reinvented itself as a biotechnology company in 1976 and hired Stanley Cohen to head its new research center in Palo Alto. Further funding for Cetus came from companies interested in the new technology.

A consortium of Japanese companies owned 1.59% of Cetus. Standard Oil of Indiana purchased 28% of their stock. Standard Oil of California bought 22.4%. National Distillers & Chemical purchased 14.2%, and other corporate investors who wanted to learn about the new technology also purchased stock.

Both Genentech and Cetus offered stock-options to their key scientists. Genentech and Cetus were the first of the biotechnology firms to go public. Genentech realized net proceeds of $36 million. At the end of fiscal year 1981, it had $30 million cash, but required about a million yearly to fund its R&D activities.

In its public offering, Cetus raised $115 million at $23 a share. Of this, $27 million was intended for the production and distribution of Cetus-developed product processes, $25 million to fund the R&D, $24 million for research administrative facilities, $19 million for additional machinery and equipment, and $12 million to finance a string of new-venture subsidiaries.

For new firms, it is important to come up with early products to generate income. Back in 1982, Genentech's product interests were in health care, industrial catalysis, and agriculture. In 1982, their early products included genetically engineered human insulin, human growth hormone, leukocyte and fibroblast interferon, immune interferon, bovine and porcine growth hormones for foot-and-mouth vaccine. Genentech's human insulin project was a joint venture with Eli Lilly, aimed at a world market of $300 million in animal insulin. Genentech's human growth hormone project was a venture with KabiGen (a Swedish pharmaceutical manufacturer), with a world market of $100 million annually. The leukocyte and fibroblast interferon was a joint venture with Hoffmann-La Roche, and the immune interferon with Daiichi Seiyaku and Toray Industries. The bovine and porcine growth hormones were a joint venture with Monsanto, and the foot-and-mouth vaccine with International Minerals and Chemicals.

In 1982, Cetus was interested in products in health care, chemicals, food, agribusiness, and energy. Their commercial projects included high-purity fructose, veterinary products, and human interferon. The high-purity fructose project was a joint venture with Standard Oil of California. By 1983, Cetus introduced its first genetically engineered product, a vaccine to prevent scours, a toxic diarrhea in newborn pigs. However, product income then was not rising fast enough for Cetus. In 1983, a new president, Robert A. Fildes was recruited to lead Cetus. He developed a five-year strategy to refocus Cetus on just two major health care markets, diagnostics and cancer therapeutics. Both monoclonal antibody and recombinant DNA technologies were employed. Also he reduced employment to 480 people (with 280 scientists) to develop and market drugs.

However, the path to success for the new biotechnology industry of the late twentieth century did not go smoothly. The reason was that the complexity of nature turned out to be more than what had been anticipated by the new biotechnology industry. This was hard on the profitability of the nascent industry. In 1996, Thayer commented on the industry: "Fighting waves of hype and pessimism — while trying to create products and access markets — has tested (biotechnology) firms' ability to endure". (Thayer, 1966)

By 1996, the new biotechnology industry had created 35 therapeutic products, which then had a total annual sale of over $7 billion dollars. These biopharmaceutical products were used to treat cancer, multiple sclerosis, anemia, growth deficiencies, diabetes, AIDS, hepatitis, heart attack, hemophilia, cystic fibrosis, and some rare genetic diseases. But the industry had not been as successful as early investors had hoped for.

Genentech hoped that a producing a protein product called TPA would catapult them into the large firm status. But in 1990, the costs of developing products and the relatively small market for TPA placed Genentech in financial difficulties. To survive, Genentech sold 60% of its equity to Hoffman-La Roche: "Despite TPA's success today, it took the 20-year-old company many years and many millions of dollars to prove that it had an important product." (Thayer, 1966)

In the 1990s, most of the marketing of the new biotechnology therapeutic products went through older pharmaceutical firms, rather than the new biotechnology firms, which had pioneered the recombinant DNA technology. Genentech had partnered with Hoffman-La Roche, and Chiron with Ciba-Geigy. Other start-ups from the early 1980s were firms such as Biogen, Amgen, Chiron, Genetics Institute, Genzyme, and Immunex. Biogen did pioneering work on proteins such as alpha-interferon, insulin, and hepatitis. But to survive, Biogen licensed their discoveries so that other pharmaceutical firms could market their products. Genetics Institute and Immunex were majority owned by American Home Products. One exception to this pattern describing the fate of early biotechnology firms was Amgen. In 1995, Amgen had become an industry leader in biotechnology — an independent, fully integrated biopharmaceutical producer with sales of $1.82 billion.

Why had the early dream-for-big-success in biotechnology not come true? The answer to this question was in the 'complexity' of biology: "Early expectations, which in hindsight could be considered naive, were that drugs based on natural proteins would be easier and faster to develop... However... Biology was more complex than anticipated." (Thayer, 1996)

For example, one of the first natural proteins, alpha-interferon, took ten years to gain acceptance in antiviral therapy. When interferon was first produced, there had not been enough available to really understand its biological functions. The production of alpha-interferon in quantity through biotechnology techniques finally allowed real studies and experiments to be undertaken to study its therapeutic impact. This combination of developments in technology that allowed the production of proteins in quantity for therapeutic applications took a long time and cost millions of dollars.

By 1992, Cetus had spent millions of dollars and bet everything on interleukin-2 as an anti-cancer drug; but it failed to obtain the United States Federal Drug Administration's (FDA) approval to market the drug for this purpose. Subsequently, in

1992, Chiron acquired Cetus. Even in 1995, interleukin-2 (which was eventually approved by the FDA) contributed only 4% of Chiron's revenues of $1.1 billion. About this, George B. Rathmann commented: "...the pain of trying to get interleukin-2 through the clinic just about bankrupted Cetus and never has generated significant sales." (Thayer, 1996)

The innovation process for the biotechnology industry in the United States include

(1) developing a product,
(2) creating a production process,
(3) testing the product for therapeutic purposes,
(4) submitting the product to the US's FDA for approval of its usefulness and safety, and
(5) marketing the product.

The recombinant-DNA technique was only a small part of the technology needed by the biotechnology company (and the smallest part of its innovation expenditures). The process of testing the product to gain FDA approval took the longest time (typically seven years) and required a very deep pocket.

Because of this long and expensive FDA process in the U.S., extensive partnering occurred between U.S. biotech firms and the larger, well-established pharmaceutical firms. In 1995, pharmaceutical companies spent $3.5 billion to acquire biotechnology companies and $1.6 billion on R&D licensing agreements (Abelson, 1996). Pharmaceutical firms spent more than $700 million to obtain access to data banks on the human genome that was being developed by nine biotechnology firms.

The U.S. Government's support of science was also essential to the U.S. Biotechnology industry according to Henri Termeer, chairman and CEO of Genzyme and chairman of the U.S. Biotechnology Industry Organization: "The government has a very big role to play (in helping) to decrease the costs. Support of basic research through NIH (National Institutes of Health) is very important to continue the flow of technology platforms on which new breakthrough developments can be based." (Thayer, 1996)

The scientific importance of understanding the molecular nature of biology (the discipline now known as 'molecular biology') did prove to be the future of the pharmaceutical industry — as an essential methodology to develop new drugs. Yet the making of money from the technology of recombinant DNA was harder and took longer than expected because biology turned out to be more complicated than anticipated. The biotechnology industry's technology depended on and will continue to depend on new science. In turn, the technology needs of the biotechnology industry have helped to drive the discoveries in biological science. *The progress of a new technology depends on the progress of science in understanding the complexity of nature underlying the technology.*

Nevertheless, all of the expense going into the biotechnology industry did turn out to be worthwhile. Molecular biology continues to be the future of the pharmaceutical industry. For example, in 1996, George Rathmann summarized the situation nicely: "It doesn't have to follow that science automatically translates into great practical results, but so far the hallmark of biotechnology is very good science and that now is stronger and broader than ever... The power of the science is ample justification that there should be good things ahead for biotechnology." (Thayer, 1996)

Research and Complexity of Nature

Thus, the application of biotechnology to medicine and to agriculture had not immediately created great wealth at the beginning for the industry. This was because biology is complex. Its nature was not well enough understood, yet it needed to be understood so that effective and safe medical cures or agricultural applications could be developed. The subtle complexities of nature are the sources of all the technological and engineering problems with physical/biological technologies. *The more complex nature (which underlies a technology) is, the more scientific research is needed — to further understand the nature behind the technology.*

Even though a scientific phenomenon has been discovered from which a new technology could emerge, which could in turn give rise to a new invention — there is still a long way to go before a product can be developed that performs effectively and efficiently at a reasonable cost. This step, which comes after science and technology research, is known as 'engineering research' or product 'development'. Engineering research further develops the technology to operate at the performance specification required by a product. Engineering research deepens and fills in the knowledge base of science that is necessary to improve a technological invention. A technology must be improved sufficiently to enable it to manipulate a natural phenomenon well enough that the resulting invention performs at a level which makes it useful to mankind. The more complex nature of biology has made the science research and the engineering research indistinguishable. As in biotechnology, engineering progress moves at the pace of scientific progress and vice-versa.

Historical Case: Human Genome Project

The complexity of biology is that science needs to discover both the biological 'mechanism' and biological 'function' of a biological component. A 'Mechanism' is how nature operates mechanically, and a 'Function' is the value of mechanistic processes to humanity.

The next big molecular biology project, the 'human genome' project, was undertaken at the end of the twentieth century. Its aim was for researchers to write down and decode the set of instructions for the replication of an entire human being. To sequence the human genome, one must place in order three billion chemical units, and these encode the instructions on how to build the biological units of a human organism. In the 1990s, the human genome project was pursued by two independent and competing scientific groups.

On the one hand, one group was an international public consortium consisting of the National Centre for Biotechnology in the United States, the European Bioinformatics Institute and the DNA Data Bank in Japan. This group was funded for 15 years with a total of about $3 billion dollars. On the other hand, the second competing group (operating on private industrial funding) was the company of Celera Genomics in Rockville, Maryland, U.S.A. In June 2000, both groups announced a first draft of the chemical sequences in all the DNA of the human genome.

The mechanism of transmitting heredity in animate objects (living organisms) is DNA replication. The two strands of the DNA molecule are composed of four nucleotide units, or bases, each made up of three amino acids — adenine, thymine, guanine and cytosine (A, T, G, C). These bases are complementary pairs, with adenine (A) complementary to thymine (T), and cytosine (C) complementary to guanine (G). This base pairing in chemical attraction is the physical mechanism for structuring the DNA molecule and its complementary strands. It is also the mechanism for reproducing an organism's cells and for creating proteins to be used in the operations of an organism. The two stands of DNA helix separate to form single-stranded DNA. For each strand, adenine (A) attracts thymine (T), and guanine (G) attracts cytosine (C) to direct the synthesis of two new double-helixes. In this way, cells copy their DNA before dividing into daughter cells.

The success of the human genome project depended on scientific technology for automating the sequencing of DNA. The scientific technique of DNA sequencing, polymerase chain reaction (PCR), was invented in 1977 by Frederick Sanger at Cambridge University in the United Kingdom. DNA sequencing begins with the preparation and replication of a segment of

DNA to be sequenced. Next, many partial, but shorter fragments of the segment are also replicated, each part with one less base than the prior part. Then the ending base (A, T, G, or C) is tagged with a fluorescent chemical. These fragments are all separated according to length, and the length and ending base of each is recorded. Finally, the sequence of the bases in the original DNA segment can be reconstructed by placing the different lengths of the DNA fragments in order and identifying the ending base of each. The DNA segment sequence is the sequence of ending bases arranged by the length of the replicated fragments.

In 1989, the sequencing of DNA was not yet automated, and required a great deal of technician labour, with a cost of about $2 to $5 dollars for identifying a base in a DNA sequence. Automation was essential as the planned budget of $15 billion for the U.S. Human Genome project had been formulated at an estimate of $0.50 dollars per base identification. Then Daniel Cohen, who was the director of a scientific laboratory, Genethon, in Paris, France, began to make significant progress in automation. Cohen implemented 18 machines, each of which could output the identification of 8000 bases per day. He designed custom robots to place 450,000 samples of DNA per day on a membrane for reactions. Genethon published its physical map of the human genome in 1993 as an ordered set of large blocks of DNA (yet un-sequenced within a block). This was used to break the human genome down into a set of blocks.

The progress of the DNA sequencing machinery for automating the sequencing technique advanced rapidly. In the year 2000, a DNA sequencing machine could automatically produce 330,000 bases per day. Sequencing machines prepare and replicate short segments of DNA. Thereafter, these segments are ordered by computer systems assembled in the form of a cluster of computer working in a group. This is an example of engineering research in biotechnology — the automation of the sequencing process. *Here one can here see a circular interaction between science and technology and engineering; science research created a market for the technology of sequencing and engineering research automated the sequencing technology and made possible the scientific project of decoding the human genome.*

Once a fragment of a DNA sequence was automatically analysed according to the order of its bases, a computer program could then assemble these bases into the DNA sequence. In 2000, Phil Green and his colleagues at the University of Washington Genome Centre in Seattle, Washington, USA, had developed assembly software to automate the sequencing process. One program, Pred (Phil's read editor), looked at the fluorescent signal output from the DNA sequencers, identified the base according to the fluorescent signal — and analysed the quality of the signal to estimate the accuracy of the identification of the base. Another computer program, Phap (for Phil's revised assembly program), assembled the DNA segment sequence by overlaying the fragment pieces to get an end-to-end sequence of the segment. A third program, Consed (consensus visualization and editing program) allowed a human finisher to see and edit the sequence of the DNA for storing in a DNA library of the genome.

The new technology for automatically sequencing DNA made possible the historic race between the human genome project and Celera. There were about 30,000 human genes. Still, this was only a first draft of the "book of the human genome" — only the first step.

Sequencing DNA was a first step to understanding how DNA controls the construction and functions of humans. Next, genes within each chromosome need to be identified, as genes are the functional aspect of DNA. Once genes are identified, their functionality can be established. This was and continues to be very complex research, which is still underway today.

Cyclic Interaction between Science and Technology

The human genome project and the automation of gene analysis techniques showed how important instruments are to scientific progress, for instance, in the development of technology from science. Thus, in addition to the radical innovation process (from nature to utility), there is also an incremental innovation process (from science to technology) that leads to scientific instrumentation. Perpetuating this incremental process, which connects science and technology, is important and challenging because the information processes in science and in technology follow two different knowledge tracks: *progress in science* and *innovation in technology*.

The scientific knowledge needed to progress in science can be found in universities, in which the scientific knowledge base provides the content for the university science and engineering disciplines. This leads university researchers to ask more fundamental questions about nature, from which they formulate research projects.

The technological knowledge needed to create technological innovations can be found in industries, where industrial R&D carries out research projects to improve their products/services. These projects enable the innovation and design of new products and production processes.

These two knowledge tracks are separate and need to be coordinated for science to feed technology and vice-versa. This is not easy and is, in fact, the major source of failure for government R&D investments, which do not create innovation and national competitiveness. This failure happens when knowledge generated in the science track is not coordinated with knowledge generated in the technology track.

Thus 'research cooperation', to connect the science track with the technology track, is the critical policy element in S&T policy for a successful national system. Research cooperation is difficult because there are vast institutional gaps between the three sectors; namely universities, industry, and government. Universities specialize in scientific progress, while industry specializes in technological progress. The government specializes in putting in place policy for funding science and military technology. Institutionally, all their respective missions differ greatly, as do their goals, timing, domains, finances, and

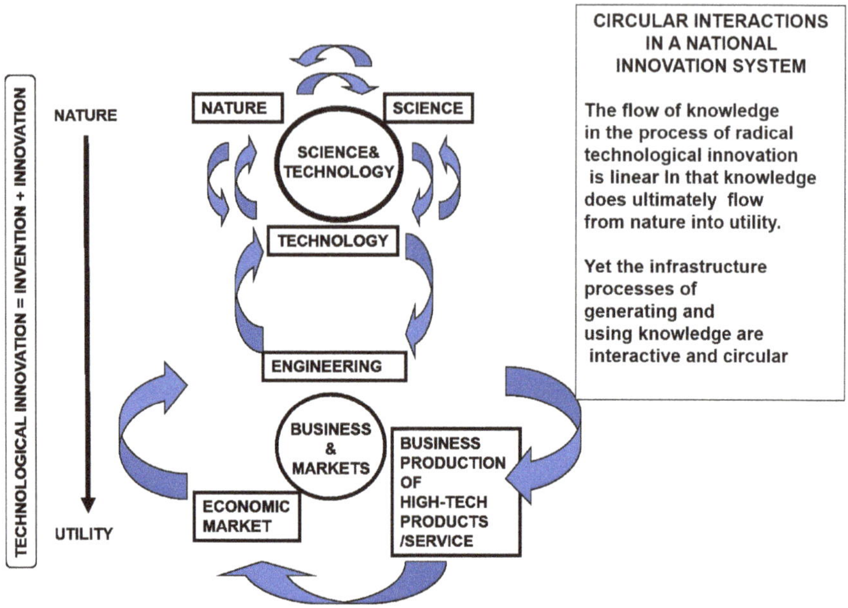

Figure 3.2. The circular interactions between knowledge and radical innovation.

cultures. Nevertheless, the benefits for cooperation are enormous, with modern examples including the university/industrial cooperation in inventing and developing computers, biotechnology and the Internet.

While the flow of knowledge is linear in the innovation process from nature to utility, in Science & Technology, the interactions between each component of science, technology and engineering could be circular.

Summary

In this chapter, we have looked at the origin of the biotech industry, and the progress it has made in understanding DNA sequences in the human genome project. We have also looked at how science can be a basis for developing technology and inventing products that can manipulate nature for human utility. This process requires research to work hand-in-hand within the three R&D sectors of university, industry, and government. While the innovation process from nature to utility is linear in nature in terms of the contribution to knowledge made at each stage, the process could be circular in the interactions between science, technology, and economy.

CHAPTER 4

Targeting Basic Research

Introduction

We have emphasized that science is expensive, and that the government should be the principal sponsor of science. But what types of science should governments sponsor? One answer is that governments should fund science which could help to deliver the objectives underpinning the goals set by the government concerned. Government research agencies are mostly organized into government-mission-focused agencies, such as departments of defence, agriculture, medicine, commerce, transportation, etc. Each government-mission-focused agency can support research on technology & science relevant to the mission set by the government. Let us now look at how a government agency formulates a funding strategy to accomplish their mission and deliver goals on the government's behalf.

In the United States, the expansion of government support for research took place mainly after the Second World War in the twentieth century. Science & technology policies by the U.S. government expanded; resulting in an increase in government funding for research in American universities. In particular, significant levels of research were funded in the Department of Defence, National Institutes of Health, National Science Foundation, and Department of Agriculture. In the second half of the twentieth century, government funding for research in the U.S. overwhelmed funding given by private philanthropy in universities.

Case History: U.S. National Institutes of Health

We will review a case of research strategy in the U.S. National Institutes of Health (NIH) formulated in 2007. Each year, U.S. government agencies must submit strategies and desired budgets to the U.S. Congress for funding. Back in 2007, the impacts of molecular biology and genome mapping were so great that NIH refocused their strategy to the then new scientific challenges in biological science, underpinning medicine, to formulate a comprehensive strategy to develop biological science. We will look at this strategy from the perspective of the thinking that led to the realization of this objective.

The U.S. federal government's responsibility for advancing medical research through R&D is embodied in the NIH agency within the U.S. Department of Health and Human Services. The main campus of its institutes is in Bethesda, Maryland, just on the northern border of the capital of the U.S. in Washington, D.C. The mission of NIH is "science in pursuit of fundamental knowledge about the nature and behaviour of living systems and the application of that knowledge to extend healthy life and reduce the burdens of illness and disability." (NIH[a], 2007) NIH funds research both within its own governmental research centres (institutes) and externally for U.S. universities, hospitals and small businesses.

In 2007, NIH identified its primary goals as the fostering of discoveries in biology and the application of these discoveries to medical and health care services. To achieve these goals, NIH supported research focused on:

- the causes, diagnosis, prevention, and cure of human diseases;
- the processes of human growth and development;
- the biological effects of environmental contaminants;
- the understanding of mental, addictive and physical disorders;
- directing programs for the collection, dissemination, and exchange of information in medicine and health" (NIH[a], 2007)

From this list we see that NIH first focused on public welfare ("causes, processes, biological effects, understanding of disorders") and then on research infrastructure for health ("collection, dissemination, and exchange of information in medicine and health").

The history of the NIH began in 1798, when the government of the new nation of the U.S.A. established a Marine Hospital Service for seamen. In 1836, the U.S. Army established a Library of the Office of Surgeon General; and in 1887, the government established a Laboratory of Hygiene (to deal with infectious diseases). A National Advisory Health Council was established in 1902, with the name of the Marine Hospital Services changing to Public Health and Marine Hospital Services (later becoming the Public Health Service in 1912). In 1930, the Hygienic Laboratory was renamed the National Institute of Health (NIH). The National Cancer Institute was established in 1937. In 1946, a Division of Research Grants was added within NIH to fund external medical research. Many more new institutes for different health problems were established from 1948 through 1997. By 2007, NIH had numerous institutes and centres, with a total budget of 28 billion dollars and over 18,700 employees. The institutes could be grouped into four categories according to research focus: those whose research focused on health problems, those whose research focused on systems, those whose research focused on biological nature, and those whose research focused on medical technology.

NIH Institutes

Health Problems

National Cancer Institute (NCI)
National Eye Institute (NEI)
National Heart, Lung, and Blood Institute (NHLBI)
National Institute on Alcohol Abuse and Alcoholism (NIAAA)
National Institute of Allergy and Infectious Diseases (NIAID)

National Institute of Arthritis and Musculoskeletal and Skin Diseases (NIAMS)
National Institute on Deafness and Other Communication Disorders (NIDCD)
National Institute of Dental and Craniofacial Research (NIDCR)
National Institute of Diabetes and Digestive and Kidney Diseases (NIDDK)
National Institute on Drug Abuse (NIDA)
National Institute of Mental Health (NIMH)
National Institute of Neurological Disorders and Stroke (NINDS)

Systems — Biological, Environmental, Institutional

National Institute of Child Health and Human Development (NICHD)
National Institute on Aging (NIA)
National Institute of Environmental Health Sciences (NIEHS)
National Institute of General Medical Sciences (NIGMS)
National Institute of Nursing Research (NINR)

Biological Nature

National Human Genome Research Institute (NHGRI)

Medical Technology

National Institute of Biomedical Imaging and Bioengineering (NIBIB)

(NIH[a], 2007)

Focus of Government Research Programs

One can generalize from the classification of NIH institutes that there are four important criteria to guide research strategies in government programs: research focused on problems, systems, nature or technology.

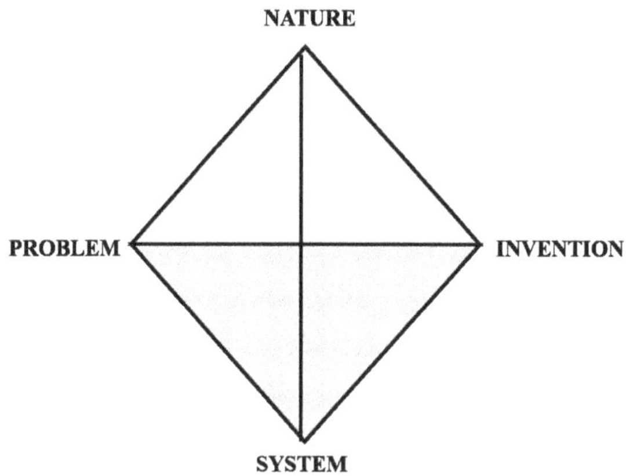

Figure 4.1. Dimensions of research strategy.

Figure 4.1 shows how planning in government research programs can be explained by the different research policies for a government mission:

(1) Focus on research about a problem in a society,
(2) Focus on research to invent and improve technology to solve the problems,
(3) Focus on the nature (science) underpinning the societal problems,
(4) Focus on representation or realization of systems (engineering) operating to solve societal problems.

In the list of NIH institutes stated earlier, we can see how NIH and their research institutes have evolved with foci on biological nature, health problems, invention of medical technology, and biological and environmental systems, as shown in Figure 4.2.

Government Research Programs

In general, government research funding focuses on these four missions by supporting (1) national research infrastructure,

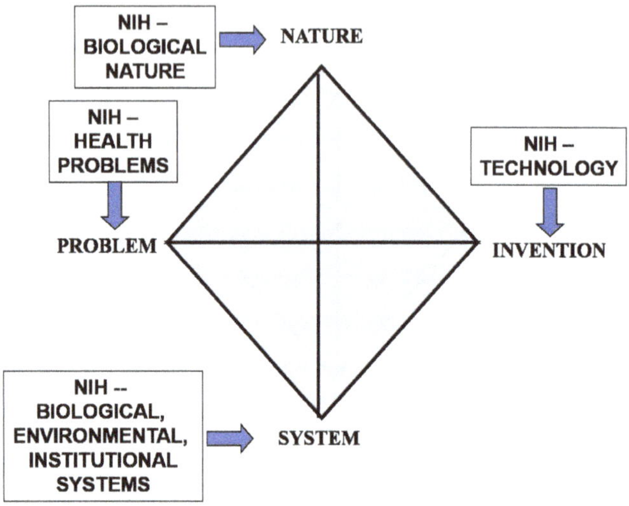

Figure 4.2. The research strategy for government programs.

(2) government functions, (3) areas of public welfare, or (4) national economic development.

For example, military force is a function of national governments; hence research and laboratories focused on weaponry technology can target the government as a customer. Since a national defence industry develops to commercialize the technology as weapon systems, some of the weapons may be sold to other friendly nations. Another example would be research in medicine (such as in the U.S. NIH), which contributes to public welfare and national infrastructure for public health. Government funding for science in university research (such as in the U.S. NSF) contributes to the national educational infrastructure and public welfare when scientific progress advances technology.

Shown in Figure 4.3, government science & technology issues can focus on providing research funding for developing (and maintaining)

(1) a nation's knowledge infrastructure,
(2) technology for use in government functions,

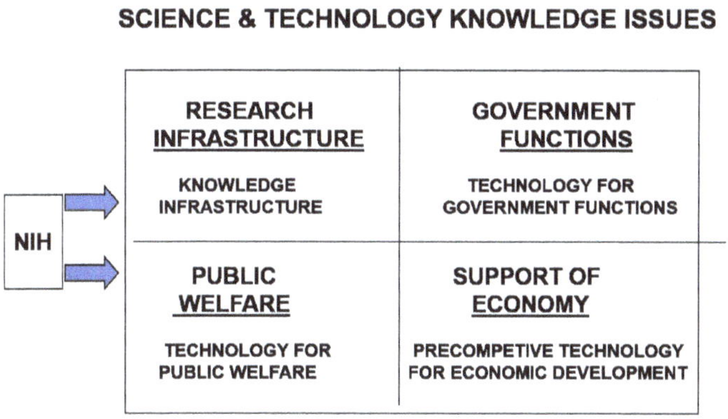

Figure 4.3. The issues science & technology knowledge.

(3) technology for use in areas of public welfare, or
(4) technologies prior to use by specific industries (pre-competitive technologies).

We can see that the NIH mission had focused on the aspects of:

(1) supporting research infrastructure in medical research, and
(2) developing medical technology for public welfare.

Research Planning in NIH

In 2007, NIH planned for the next fiscal budget year in terms of strategic program initiatives, which NIH called a "Roadmap for Medical Research". Underpinning the roadmap was the idea of the human body as a system: "The human body is dauntingly complex. To truly revolutionize medicine and improve human health, we need a more detailed understanding of the vast networks of molecules that make up our cells and tissues, their interactions, and their regulation. We also must have a more in-depth knowledge of the combination of molecular events leading to disease." (NIH[a], 2007)

Often, science progresses until major breakthroughs occur. Research issues can remain current for many years. Even though this case of research strategy occurred in 2007, many of the basic issues then still remain current. This issue of depicting the human body as a system remains an issue even now, a decade later in 2017.

Strategic research planning must not only express the focus of the research (the human body) but also find scientific opportunities to guide the research focus. NIH envisioned the scientific opportunities as arising in molecular biology: "To capitalize on the completion of the human genome sequence and recent exciting discoveries in molecular and cell biology, the research community needs wide access to technologies, databases, and other scientific resources that are more sensitive, more robust, and more easily adaptable to researchers' individual needs." (NIH[a], 2007) With a focus on the system of the human body and research opportunities in molecular biology, NIH identified five research initiatives in 2017:

1. BUILDING BLOCKS, BIOLOGICAL PATHWAYS, AND NETWORKS
2. MOLECULAR LIBRARIES & MOLECULAR IMAGING
3. STRUCTURAL BIOLOGY
4. BIOINFORMATICS AND COMPUTATIONAL BIOLOGY
5. NANOMEDICINE

(NIH[a], 2007)

Building Blocks, Biological Pathways, and Networks

"Complex elements — from individual genes to entire organs — work together in a feat of biological teamwork to promote normal development and sustain health. These systems work because of intricate and interconnected pathways that enable communication among genes, molecules, and cells. Scientists are still working to discover all these pathways and to determine how disturbances in them may lead to disease.... Ultimately, scientists hope to be able to completely map an organism's protein and metabolism networks, and create models to help predict the human body's response to disease, injury, or infection."

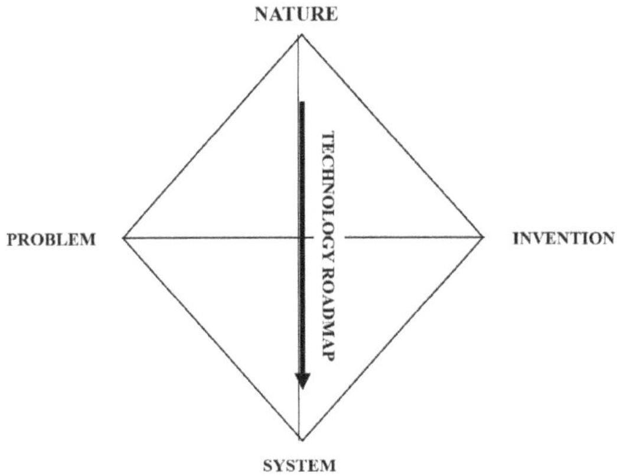

Figure 4.4. The technology roadmap from nature to system.

(NIH, 2007) *In Figure 4.4, we see that this NIH technology research roadmap begins with funding research on biology to better understand natural systems.*

Molecular Libraries & Molecular Imaging

"The Molecular Libraries initiative will provide public sector biomedical researchers access to small organic molecules that can be used as chemical probes to study the functions of genes, cells, and biochemical pathways in health and disease. The component is also expected to facilitate the development of new drugs by providing early-stage chemical compounds to researchers for them to find successful matches between a chemical and its target, and thus, help validate new targets with potential for therapeutic intervention. The ultimate goal is to enable a detailed molecular understanding of cell and tissue function in normal and disease states, which may lead to greater power to diagnose and treat disease." (NIH[a], 2007) *In Figure 4.5, we see that the NIH technology research roadmap begins with funding research on biology to invent new medical treatments (therapeutic interventions).*

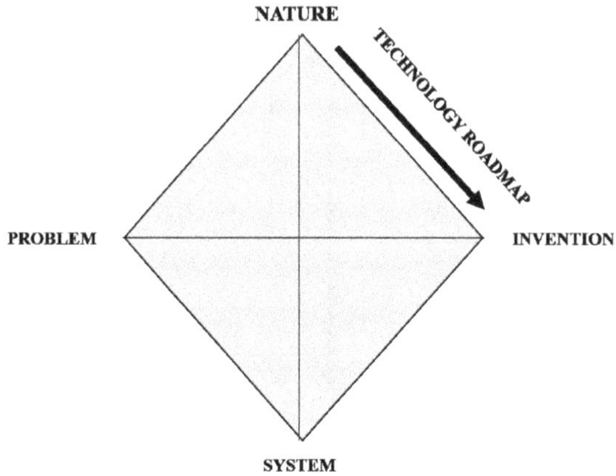

Figure 4.5. The technology roadmap from nature to invention.

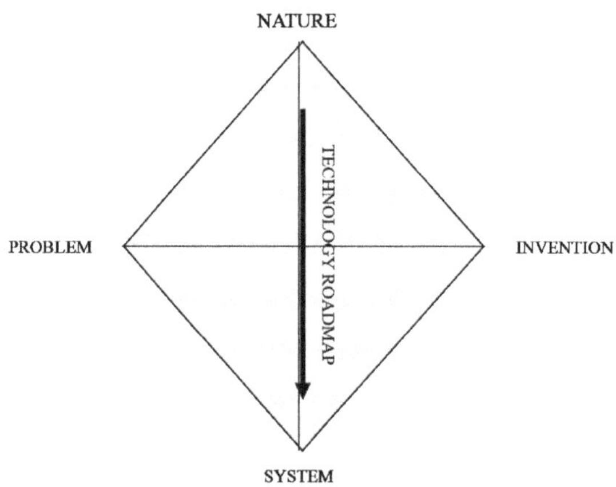

Figure 4.6. The technology roadmap from nature to system.

Structural Biology

"Proteins are indispensable molecules in our bodies, and each has a unique three-dimensional shape that is well-suited for its particular job. . . This initiative is an effort to create a picture gallery of the molecular shapes of proteins . . . to advance our understanding of how proteins and their component parts function in the body." (NIH[a], 2007) *In Figure 4.6, we see this NIH technology research roadmap begins with funding research on biology to better understand natural systems.*

Bioinformatics and Computational Biology

"Scientists are using computers and robots to separate molecules in solution, read genetic information, reveal the three-dimensional shapes of proteins, and take pictures of the brain in action. These techniques generate huge amounts of data, and biology is changing fast into a science of information management. Researchers need software programs and other tools to analyse, integrate, visualise, and model these data." (NIH[a], 2007) *In Figure 4.7, we see this NIH technology research roadmap begins with funding research for instrumental inventions to improve the study and modelling of biology.*

Nanomedicine

"A long-term goal is to create materials and devices at the level of molecules and atoms to cure disease or repair damaged tissues, such as bone, muscle, or nerve..." (NIH[a], 2007). *In Figure 4.8, we*

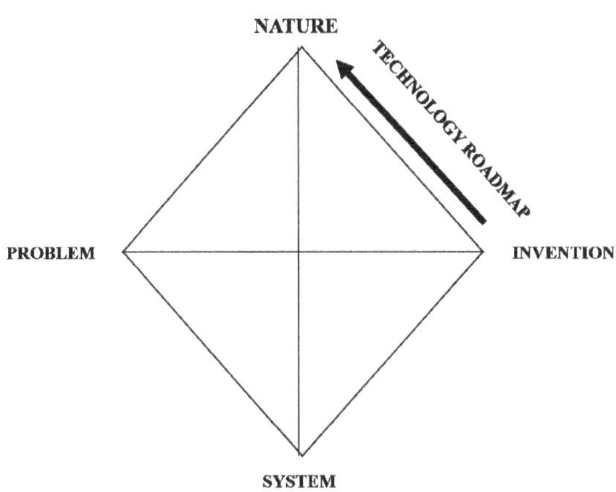

Figure 4.7. The technology roadmap from invention to nature.

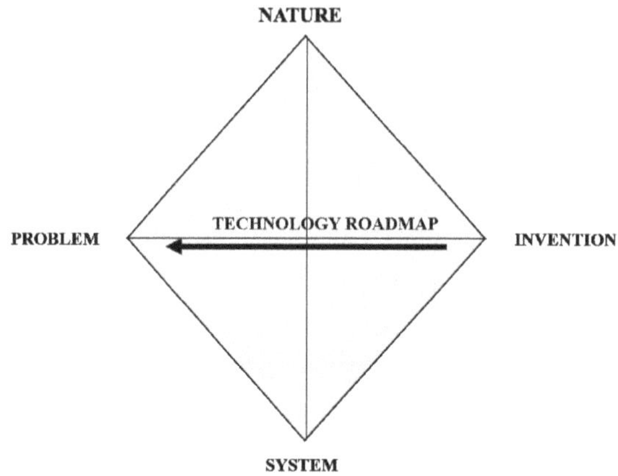

Figure 4.8. The technology roadmap from invention to societal problem.

see this NIH technology research roadmap begins with funding research for technological inventions to improve medical solutions to health problems.

Science 'Roadmaps' in Research Planning

In the NIH example, we saw a case of research program planning based primarily upon science, and this was seen in the directions of the connections from nature to system, nature to problems or technology to nature. NIH used the term 'technology-roadmap' for their planning, but a more accurate term should have been 'science roadmap'. We saw that the mapping was directional to and from nature and problems-systems-technology. *Research program strategic initiatives can be formulated through connecting research in science to research in technology to research in problems. We summarize this strategy approach in* Figure 4.9.

The strategic program initiatives create connections between the strategic dimensions in the following ways, which correspond to the numbered arrows in Figure 4.9:

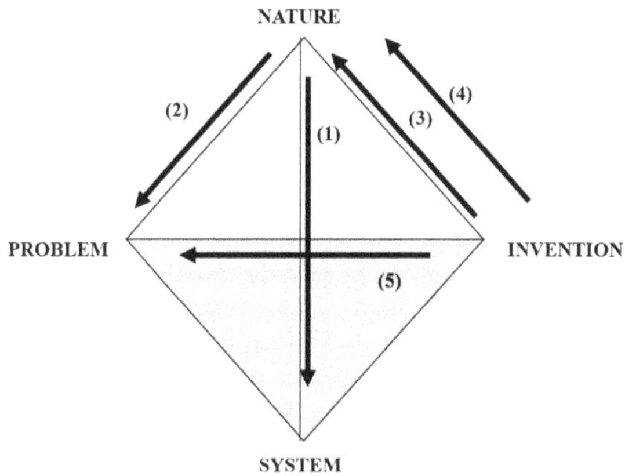

Figure 4.9. The technology roadmaps from one strategic dimension to another.

- Progress in studying nature leads to improved systems representation (1)
- Progress in studying nature leads to improved problem analysis (2)
- The invention of new technologies leads to improved instrumentation (3)
- The invention of new technologies leads to improved scientific techniques (4)
- The invention of new technologies leads to improved problem solving (5)

'Science roadmaps' can facilitate the planning of research initiatives based on scientific opportunities, hence improving the understanding of nature underlying societal problems and applications.

Peer Review of Quality of Research Proposals

Government research budgets determine the areas (and levels of support) for scientific research to be undertaken in the nation. How these budgets are allocated to individual research projects will be next determined at the macro-level — by the research-project selection procedures of a government agency. For quality in research, project selection should occur through a peer-reviewed, methodological procedure. In any public research funding program — such as in the US's NSF or Turkey's TUBITAK or the European Union's European Research Commission (ERC) — selection criteria are publicly stated and then adhered to in the review procedure.

Generally speaking, in government research grant programs, the principal criteria for scientific quality are merit and significance. For example, the U.S. NSF official peer review criteria are:

Criterion 1: What is the intellectual merit of the proposed activity?

- "How important is the proposed activity to advancing knowledge and understanding within its own field or across different fields?
- How well qualified is the proposer (individual or team) to conduct the project? (If appropriate, the reviewer will comment on the quality of prior work.)
- To what extent does the proposed activity suggest and explore creative and original concepts?
- How well conceived and organized is the proposed activity? Is there sufficient access to resources?"

Criterion 2: What are the broader impacts of the proposed activity?

- "How well does the activity advance discovery and understanding while promoting teaching, training, and learning?
- How well does the proposed activity broaden the participation of underrepresented groups (e.g., gender, ethnicity, disability, geographic, etc.)?
- To what extent will it enhance infrastructure for research and education, such as facilities, instrumentation, networks, and partnerships?

- Will the results be disseminated broadly to enhance scientific and technological understanding?
- What may be the benefits of the proposed activity to society?"

(NSF 99-172, OFFICE OF THE DIRECTOR, September 20, 1999, National Science Foundation Merit Review)

Turkey's TUBITAK also used the following three criteria for peer reviewing the selection of proposals in 2007:

- Intellectual/Scientific/Professional Merit
  Theoretically sound, Novel, and possibly Significant Contribution
- Achievability of the Research
  Sound method, Appropriate and Complete team, Proper equipment
- Expected Impact of Anticipated Outcome
  In Science or in Technology or in Industry or in Society.

(TUBITAK, 2007)

The quality and range of science in a nation are strongly influenced by the procedures of research administration — at both the macro and micro levels of science. The macro-level of the organization of research is focused on selecting and funding research. The micro-level of the organization of research is focused on gaining funding, performing and using research.

The macro-level quality of science & technology policy is at the heart of modern economic development. Over the last two hundred years, economically competitive nations have been those that advanced in science/technology. Why are some nations in the world today advanced in science and technology, while others are lagging? The answer lies in the differences between nations in:

(1) *The quality of institutional procedures of science,*
(2) *The quantity of funding received by science & technology research,*

(3) *The national capability to apply science & technology by transforming them into economic utility.*

ILLUSTRATION: European Union Technology Platforms in 2006–2007

In addition to 'science roadmaps', research planning for technology has used the concept of 'technology roadmaps' to formulate strategic research initiatives based on technology. An example of the technology-based approach was the European Union (EU) Framework Program in 2006–2007: "Framework programs have been the main financial tools through which the European Union supports research and development activities covering almost all scientific disciplines... The seventh Framework program runs for seven years (2007–2013)... with a budget of 50 million euros." (CORDIS[a], 2007)

The EU Framework programs described the research areas to be supported in terms of a technology platform: "European Technology Platforms (ETPs) provide a framework for stakeholders, led by industry, to define research and development priorities, timeframes and action plans on a number of strategically important issues, where achieving Europe's future growth, competitiveness and sustainability objectives is dependent upon major research and technological advances in the medium to long term." (CORDIS[b], 2007)

These technology platforms were planned by industrial and university researchers to identify future areas of research, in which research proposals may be submitted and are listed as follows.

- Advanced Engineering Materials and Technologies — EuMaT
- Advisory Council for Aeronautics Research in Europe — ACARE
- Embedded Computing Systems — ARTEMIS
- European Biofuels Technology Platform — Biofuels
- European Construction Technology Platform — ECTP
- European Nanoelectronics Initiative Advisory Council — ENIAC
- European Rail Research Advisory Council — ERRAC

- European Road Transport Research Advisory Council — ERTRAC
- European Space Technology Platform — ESTP
- European Steel Technology Platform — ESTEP
- European Technology Platform for the Electricity Networks of the Future — SmartGrids
- European Technology Platform for Wind Energy — TPWind
- European Technology Platform on Smart Systems Integration — EPoSS
- Food for Life — Food
- Forest based sector Technology Platform — Forestry
- Future Manufacturing Technologies — MANUFUTURE
- Future Textiles and Clothing — FTC
- Global Animal Health — GAH
- Hydrogen and Fuel Cell Platform — HFP
- Industrial Safety ETP — Industrial Safety
- Innovative Medicines for Europe — IME
- Integral Satcom Initiative — ISI
- Mobile and Wireless Communications — e-Mobility
- Nanotechnologies for Medical Applications — Nano-Medicine
- Networked and Electronic Media — NEM
- Networked European Software and Services Initiative — NESSI
- Photonics21 — Photonics
- Photovoltaics — Photovoltaics
- Plants for the Future — Plants
- Robotics — EUROP
- Sustainable Chemistry — SusChem
- Water Supply and Sanitation Technology Platform — WSSTP
- Waterborne ETP — Waterborne
- Zero Emission Fossil Fuel Power Plants — ZEP

(CORDIS[b], 2007)

Program Planning and Evaluation

To evaluate a government research funding program, one should compare performance against plans. Periodically, government research programs can be reviewed to judge the effectiveness of a program's past support and to guide strategic decisions on future directions and budget. But such program reviews of performance need to use prior program plans against which to judge performance. If no prior research plan is in place, future plans cannot be carried out or evaluated; i.e. a program plan is required to evaluate program performance. For science-based research programs, such as NIH, program planning can be done with the previously mentioned tool which organizes all plans into the categories of nature, problem, system, and invention planning.

Since research programs fund research projects, the evaluation of a research program's performance also requires examining the portfolio of projects in the program and their results. However, the value of research results from a portfolio depends partly on what the program intended to accomplish as a whole.

> *All research programs must first have a strategic plan against which to measure accomplishments for effective program review.*

A program review for a research program asks the question of how many of the research goals of the research program have been attained from the program's research projects portfolio, which the program supported in prior years. This requires a database for a research program to maintain, listing research projects supported, their research results and the use of those results.

When a research program supports science research, then the research results in its project portfolio can be evaluated by a citation analysis of published scientific papers due to the projects. What scientific papers were published from the program's scientific research projects? How much did the relevant scientific community cite the papers? Were there any papers from that portfolio which made significant contributions to scientific progress in their fields?

When a research program supports technology research, then the research results in its project portfolio can be evaluated on the basis of contribution to technological innovation. Did anyone (government or industry) utilise any of the technology inventions, or make any improvements to the technology based on the researched science knowledge? If so, who, when and where? Were patents obtained from the technology research projects? Who used or would use the patents? Technology research can only be evaluated by its usefulness to a technology user — either a government or an industrial user.

- *Programs supporting science research can be evaluated by a citation analysis of scientific literature.*
- *Programs supporting technology research can be evaluated by the usefulness of the research to a technology user in government or industry. Government research programs can be organized around research:*
  a) *that supports the infrastructure of science,*
  b) *that addresses specific societal problems,*
  c) *that results in the invention of technologies to address societal problems.*
- *Research visions for desirable new research goals may be expressed in terms of:*
  a) *research area and*
  b) *desirable achievements from future research in the area.*

Summary

In this chapter, we have seen two terms used by NIH and EU to express research program initiatives: 'science roadmap' and 'technology platform roadmap'. Both these terms were used to indicate a direction for research programs — research goals.

A technology roadmap was meant to lay out a desired direction for technological progress toward which a research program should aim. For science-based programs such as in NIH, a more accurate term should be science roadmap. In science-based research programs, the directions in a roadmap are to and from nature (science).

The term 'technology platform roadmap' was used to identify an area of technology where technical progress would be desirable. The term 'platform' was used to indicate that technical progress in a given area of technology might have wide-ranging benefits, as many useful technologies could benefit from technical progress in that particular area. In that sense, the technical ideas underlying said progress would act as a supporting knowledge structure, or 'platform'.

Both terms are also metaphors. The term 'roadmap' indicates that progress occurs in a certain direction. Real geographical roadmaps depict routes from one place to another. But research activities are exploratory activities. Accordingly, in a research roadmap, one cannot depict a precise route or path, since it has not yet been travelled. Hence, research roadmaps should point in the direction of scientific and technological opportunities.

A 'platform' is some basis upon which to stand or build. The EU wanted research to create a base upon which to build technical knowledge, which industries could stand on and use to develop new technologies. So in their list, they had research experts from the industries and academia offer their research visions, upon which new industrial technologies might be built in the future. The core ideas in broad areas of technology can provide cognitive structures that support technical progress, which might result in many new core technical ideas that can be used in technical applications.

What is essential to both metaphors is to have a 'research vision' for desirable new technology that guides research program planning — whether as a 'science map', or as a 'technology platform map'.

CHAPTER 5

Research Cooperation: Mission Agency & Industry

Introduction

As we have witnessed in the previous chapter, the Technology Roadmap runs either from Nature to System, or from Nature to Invention. There is no denying the role played by the government in the development of science undertaken by the universities, which paves the way for industry to take advantage of discoveries in science — by inventing technology that powers the national economy. How then does science get transferred into technology, and then into the generation of profits? How does a nation deal with the intellectual property that has arisen from the national research system? These are two important issues in science & technology policy that we have to take note of: *technology transfer and intellectual property*. The answer is that government research-funding agencies should put in place policies for the transfer of technology and property to industry within cooperative research between agencies and firms.

Case History: Biotechnology Industry Progress in 2015

As we reviewed earlier, progress in biotechnology began in the second half of the twentieth century and continued into the twenty-first century. By 2015, biomolecular science had discovered a new technology for conquering cancer. Andrew Pollack (2015) wrote: "... the biotech business is experiencing an almost unprecedented boom again... Money is flowing into the industry... Stock prices are high, and drug approvals are up. And perhaps most important, some of the new drugs represent major advances against diseases like cancer, hepatitis C and cystic fibrosis. Buoyed by the recent success, some executives and investors are making the bold assertion that the industry has turned a corner and that improved scientific insights and new techniques are allowing drug developers to reduce their notoriously high failure rate and to take on some illnesses for the first time."

In 2015, a J. P. Morgan Healthcare Conference was held for four days and attended by nine thousand biotechnology investors and executives. Four hundred biotech companies presented business plans, looking for investors. Pollack (2015) wrote: "Biotechnology stocks have outperformed the overall market for several years... Over 110 companies in the industry went public last year, far more than in a typical year, raising a record $9 billion... One factor lifting stock prices is the frenzied pace at which small drug developers are being acquired by larger ones. That is in part spurred by the need for big pharmaceutical companies to replenish their pipelines as top-selling drugs go generic."

Progress at a neck-breaking speed in technology was driving the biotech industrial boom. Pollack (2015) wrote: "New hepatitis C drugs from AbbVie and Gilead Sciences can cure most patients in 12 weeks. Other drugs help the body's immune system attack cancer; two of which, approved last year (2014), have generated tremendous excitement among oncologists. Executives say new techniques for analyzing and manipulating genes and cells are taking biology to new heights, leading to the development of innovative approaches to treatment. 'The science has never been better, and the pace of progress has never been faster,' said Jeffrey Leiden, the chief executive of Vertex Pharmaceuticals..."

The important new biotechnology approach was 'cell therapy'. Pollack (2016) wrote: "In CAR-T therapy, a cancer patient's immune system cells are genetically engineered to attack tumors. There have been some substantial successes in treating leukemia and lymphoma. But the technique is still early in development and can cause severe side effects. Juno Therapeutics, a leading start-up in the field, raised $300 million last month in the industry's largest initial public offering in memory. It now has a market value of about $4.5 billion. Kite Pharma, another start-up that went public last year, has a valuation of over $3 billion. Not surprisingly, others are scrambling to get into the field."

The technique uses a patient's own immune system to attack the patient's cancer. A doctor takes immune cells from a patient, genetically alters the cells to attack the cancer cells, and injects the altered immune cells back into the patient. The altered immune cells selectively kill the cancer cells.

Pollack (2016) wrote: "The technique, known as cell therapy, gives each patient an individualized and souped-up version of their own immune system, one that 'works better than nature made it,' as Dr. June puts it. The patient's T-cells, the soldiers of the immune system, are extracted from the patient's own blood, then genetically engineered to recognize and destroy cancer. The redesigned cells are multiplied in the laboratory, and millions or billions of them can be put back into the patient's bloodstream, set loose like a vast army of tumor assassins. This is an unusual pharmaceutical — a drug that is alive and can multiply once inside the body. Dr. June calls these cells "serial killers." A single one can destroy up to 100,000 cancer cells."

These 'killer' T-cells are constructed to recognize the cancer cells and attack them. Denise Grady and Andrew Pollack (2016) wrote: "Scientists like Dr. Sadelain decided to put a new claw on the T-cells, one that could recognize cancer by latching onto a telltale protein on cancer cells. The new claws came from another part of the immune system known as antibodies. Drug companies already knew how to make antibodies with claws that bind to specific proteins in the body. But the claw was not enough. Once a claw binds to a target protein, it needs a molecule to signal the T-cell to go into killing mode. Yet another signal helps sustain the killing. The DNA instructions for all three components are inserted into the patient's T-cells... The CAR T-cells cause a ferocious immune reaction as they find and destroy the patient's B-cells, which are covered with a protein called CD19."

This new approach was different from the traditional treatment, which involved poisoning the cancer cell with externally introduced poisons or chemotherapy. Now the body itself, using immune T killer cells, can devour the cancer cells. Pollack (2016) wrote: "This radical, science-fiction-like therapy differs sharply from the more established type of immunotherapy developed by other researchers. Those off-the-shelf drugs, known as checkpoint inhibitors, release a molecular brake on the immune system, freeing it to fight the cancer much as it fights infections by bacteria or viruses. Cell therapy, in contrast, is brewed specially for each patient, one of the many challenges the field faces is in broadening its use. So far, the number of patients treated with cell therapy is in the hundreds, not thousands. And for now, it works only for certain types of blood cancers, not common malignancies like breast and lung cancer. Researchers are also still confined themselves in how to control potentially lethal side effects. Just recently, a clinical trial was briefly halted after three patients died of brain swelling."

Grady and Pollack (2016) wrote: "Bispecific antibodies are an alternative to cell therapy, one that does not require individualizing treatment for each patient. These antibodies are proteins that can attach to both a cancer cell and a T-cell, thus providing a way to bring them close together so the T-cell can attack the cancer... These drugs, known as checkpoint inhibitors, are the most widely used form of immunotherapy for cancer. They block a mechanism that cancer cells use to shut down the immune system. This frees killer T-cells — a critically important part of the immune system — to attack the tumor. Four checkpoint inhibitors have been approved by the Food and Drug Administration and are on the market; to treat advanced melanoma, Hodgkin's lymphoma and cancers of the lung, kidney, bladder, and head and neck. The drugs are also being tested in many other types of cancer."

Grady and Pollack (2016) wrote: "Though immunotherapy has been stunningly successful in some cases, it still works in only a minority of patients. Generally speaking, 20% to 40% of patients are helped by checkpoint inhibitors — although the rate can be higher among those with melanoma. Some patients with advanced disease have had remissions that have lasted for years. In some cases, combining two checkpoint inhibitors increases the effectiveness. But for some people the drugs do not work at all, or just help temporarily. Cell therapy can produce complete remissions in 25% to 90% of patients with lymphoma or leukemia, depending on the type of cancer. In some cases the remissions can last for years, but in others relapses occur within a year."

Government research policy delivered by NIH in the United States has contributed significantly to the discovery and development of the new cell therapy approach in killing cancer. NIH's policy for transferring technology to industry and working in cooperation with industry was called CRADA.

CRADA: Intellectual Property and Government-Funded Research

After the Second World War, universities and government laboratories in the United States undertook large amounts of basic scientific research, and the U. S. Government became concerned with mechanisms to transfer knowledge from science into useful technology. Since industries developed technology and universities researched science, the US government decided to transfer intellectual property from government-funded research to the university, or to the industrial company performing the research. In 1980, two legislative acts were passed to facilitate technology transfer. One was the Bayh-Dole Act (Patent and Trademark

Law Amendments Act), which allowed researchers in university or business to patent and own inventions which were funded by federal grants. Previously, the Federal Government had assumed patent rights on any inventions made with federally-funded research. In the same year, the Stevenson-Wydler Technology Innovation Act required Federal laboratories to actively participate in and fund cooperative research activities with businesses, with the understanding that businesses could patent or own inventions developed during these cooperative research activities.

A cooperative research agreement between a US government agency and a company is called a Cooperative Research and Development Agreement (CRADA). "A CRADA is a written agreement between a federal research organization and one or more federal or non-federal parties (collaborators) to work together as partners on a research project of mutual interest." (U.S. Department of the Interior, 2006)

In 2016, the National Institutes of Health posted: "CRADAs provide an exciting opportunity for NIH investigators to join with their colleagues from industry and academia in the joint pursuit of common research goals. Government scientists can leverage their own research resources, as well as serve the larger mission of NIH, to facilitate the development and commercialization of health-care pharmaceuticals and products. Companies also can leverage their own R&D efforts while collaborating in state-of-the-art NIH research. The purpose of a CRADA is to make Government facilities, intellectual property, and expertise available for collaborative interactions to further the development of scientific and technological knowledge into useful, marketable products." (NIH, n.d.)

Case History: NIH and Kite Pharma Cooperative Research in 2016

For NIH and biotechnology companies, CRADAs were important in developing the immunotherapy approach to killing cancers (killer T cells). An example was a cooperative agreement between NIH and Kite Pharma in 2016. Matt Richtel and Andrew Pollack (2016) wrote: "Enthusiasm for cancer immunotherapy is soaring, and so is Arie Belldegrun's fortune. Dr. Belldegrun, a physician, co-founded Kite Pharma, a company that could be the first to market next year with a highly anticipated new immunotherapy treatment. But even without a product, Dr. Belldegrun has struck gold. His stock in Kite is worth about $170 million. Investors have profited along with him, as the company's share price has soared to about $50 from an initial price of $17 in 2014. The results reflect widespread excitement over immunotherapy, which harnesses the body's immune system to attack cancer and has rescued some patients from near-certain death. But they also speak volumes about the value of Kite's main scientific partner: the United States government. Kite's treatment, a form of immunotherapy called CAR-T, was initially developed by a team of researchers at the National Cancer Institute, led by a longtime friend and mentor of Dr. Belldegrun. Now Kite pays several million a year to the government to support continuing research dedicated to the company's efforts."

Dr. Steven Rosenberg had led the surgery branch of the National Cancer Institute for 42 years and had been a scientific colleague of Dr. Arie Belldegrun, founder of Kite Pharma. Richtel and Pollack (2016) wrote: "(Arie Belldegrun) finished medical school in his native Israel and performed surgery in helicopters for the Israeli armed forces, completing residency at Brigham & Women's Hospital. Then Dr. Belldegrun became a research fellow for Dr. Rosenberg at the National Cancer Institute in the United States. It was 1985, and Dr. Belldegrun was put to work on a new project of Dr. Rosenberg's — extracting tumor-fighting immune cells from cancer patients, multiplying them in the laboratory, and putting them back in. 'He was one of the more outstanding fellows to come through,' said Dr. Rosenberg, 76, who is widely considered a cancer research luminary. When the fellowship ended in 1988, Dr. Belldegrun became a prominent surgeon at the University of California, Los Angeles (UCLA), but the two men stayed in touch."

While a surgeon in the medical school of UCLA, Belldegrun started a biotech company. Richtel and Pollack (2016) wrote: "Eventually, Dr. Belldegrun, 67, got the entrepreneurial bug. He co-founded a biotech company, Agensys, which was acquired by a bigger company for more than $500 million. He was also involved with Cougar Biotechnology, which developed the prostate cancer drug Zytiga and was acquired by Johnson & Johnson for $1 billion in May 2009. A month later, Dr. Belldegrun formed Kite with a group of colleagues and investors to pursue cancer immunotherapy."

With the successful sales of the start-up companies, Dr. Belldegrun had become rich; but his entrepreneurial interests continued with the new company 'Kite', focused on immunotherapy. Progress in immunotherapy was increasing. Richtel and Pollack (2016) wrote: "That same month (June 2009), a Florida marine contractor named Eric Karlson, whose non-Hodgkin's lymphoma was advancing despite four prior treatments, became the first patient treated by Dr. Rosenberg with what would eventually become KTE-C19. The treatment entailed removing some of Mr. Karlson's immune system T cells from his blood, genetically engineering them to recognize and fight his cancer, multiplying the T cells to huge numbers in the laboratory and transferring them back into his body. After two such treatments, Mr. Karlson remains alive and cancer-free eight years later."

Belldegrun learned of this success. Richtel and Pollack (2016) wrote: "... in 2010, Dr. Belldegrun visited Dr. Rosenberg and was shown the X-rays of Mr. Karlson and of a second patient. Dr. Belldegrun was bowled over. 'I had no doubt that this is going to be a drug and, more than that, it will become a platform for multiple products,' he recalled. 'We never looked back.'"

Kite then entered into a CRADA with NIH, focused upon coordinating Belldegrun's technology development in immunotherapy with Rosenberg's research in immunotherapy. Richtel and Pollack (2016) wrote: "Over the next two years, the National Cancer Institute worked out a deal with Kite that was signed in 2012. It was the first of eight contracts between the government and the company that generally take two forms.

- In one type of contract, Kite licenses patented inventions and agrees to pay the government royalties, roughly 5 percent of sales of any commercial product arising from a particular patent. However, there is no such license specifically for KTE-C19 because the underlying treatment was not patented by the N.C.I., so royalties will be minimal. Officials say the agency did not apply for a patent because the treatment was similar to what others had been developing. Also, at the time the treatment was first created, in 2007, immunotherapy was considered to have dim commercial prospects. 'Back then, we didn't even think about commercial aspects,' said Dr. James N. Kochenderfer, a scientist at the agency who designed the treatment when working in Dr. Rosenberg's group."
- By 2016, the cooperation between Kite and NIH had increased to the level that Kite was funding additional research at NIH by Rosenberg. Richtel and Pollack (2016) wrote: "Under the second type of contract, known as a cooperative research and development agreement, Kite provides money to the N.C.I. to support research. Kite is now paying $3 million a year to Dr. Rosenberg's lab and has provided $7.5 million to it in total since 2012. Based on its regulatory filings, Kite is paying $7.8 million a year for research agreements and licenses in total, with at least $4 million of that going to the cancer institute and the rest to academic or corporate partners. The taxpayer has invested, too. Dr. Rosenberg estimated that the government has spent roughly $10 million over the years on what has become KTE-C19. He said Kite's $3 million a year is about equal to the taxpayer funding in that area and has helped speed research."

The research cooperation is coordinated on a bi-weekly basis. Richtel and Pollack (2016) wrote: "These days, researchers from Kite and the cancer institute, typically including

Dr. Rosenberg and Dr. Belldegrun, confer by conference call every other Thursday for 90 minutes. Kite employees have spent long periods at the N.C.I., learning how to manufacture the therapy, and how to treat patients in advance with chemotherapy. 'We shouldn't underestimate the value and the importance of N.I.H., not only to Kite but to the whole field of engineered T-cell therapy,' Dr. Belldegrun said."

The technology transfer from the government agency into industry was extensive and rapid. Richtel and Pollack (2016) wrote: "When Kite signed its first deal with the cancer agency, he said, it 'tapped into six years of monumental work that they had done.' Some immunotherapy competitors marvel at the company's coup in tapping into the agency's expertise. 'They got 20 years of research all together in one scoop,' said Dr. Carlos Paya, chief executive of Immune Design, which is pursuing a different approach. But government officials say few, if any, other companies were interested in the technology at the time Dr. Belldegrun came calling. Dr. Rosenberg said that before Kite, a few companies, including Johnson & Johnson, had looked at an earlier version of his technology but were wary because treatment involved processing each patient's cells."

Because of the value of intellectual property, the CRADA legislation was important to biotech companies, allowing them the patents or license to the patents on research developed jointly between industry and US government agencies. Richtel and Pollack (2016) wrote: "Government-developed technology available to be licensed to companies is posted on the website of the National Institutes of Health. And when the agency intends to grant a license to a particular company, it publishes that in the Federal Register, inviting public comment and possible competing offers. Both steps were taken in the case of Kite, officials said. Kite did not get everything the cancer institute has developed in the field. Some other companies, including Opus Bio and Bluebird Bio, got rights to some products, in part because the companies had special expertise that the agency's researchers desired. But Kite seems to have gotten the balance of them and N.C.I. technology accounts for the majority of its pipeline of possible products, though the company is diversifying."

While some scientists have entrepreneurial instincts, not all do. Belldegrun liked doing both research and company development; Rosenberg was only interested in research. Richtel and Pollack (2016) wrote: "Dr. Rosenberg professes no interest in the business side of the Kite relationship. He does not own stock in any company, even Kite, though he could get up to $150,000 a year in patent royalties if some of Kite's efforts pay off. Dr. Belldegrun, in contrast to his mentor, has commercial flair. He is known for his sharp business suits, lives in the Bel-Air neighborhood of Los Angeles, and seems as comfortable on Wall Street or in high society as in the operating room. Kite's relationship with the N.C.I. is an important part of its appeal to investors. In some presentations, Dr. Belldegrun has shown a photograph of himself with Dr. Rosenberg in their younger days. And he persuaded Dr. Rosenberg to speak at Kite's first big meeting for investors in June 2015."

On August 28, 2017, Gilead Sciences, a drug-maker company, announced it was buying Kite Pharma for $1.l9 billion dollars. It was paying a 29% premium above the then stock market price of Kite Pharma. Michael J. de la Merced (2017) wrote: "Gilead would acquire Kite's new cancer-fighting method, which harnesses the body's immune system to attack malignant cells. Kite's most promising treatment, for non-Hodgkin lymphoma, is expected to receive government approval within the next year." Kite Pharma issued public shares of its stock in September 2016 at $50 a share, with shares jumping to $140 a share in late August; Gilead Sciences paid $180 a share at the end of August.

The first drug approval of the new immune cell therapy approach occurred on September 1, 2017, along with the first U.S. FDA approval. Denise Grady (2017) wrote: "The Food and Drug Administration on Wednesday approved the first-ever treatment that genetically alters a patient's own cells to fight cancer, a milestone that is expected to transform treatment in the coming years. The new therapy turns a patient's cells into a 'living drug,' and trains them to recognize and attack the disease. It is part of the rapidly growing field of immunotherapy that bolsters the immune system through drugs and other therapies and has, in some cases, led to long remissions and possibly even cures. The therapy, marketed as Kymriah and made by Novartis, was approved for children and young adults for an aggressive type of leukemia — B-cell acute lymphoblastic leukemia — that has resisted standard treatment or relapsed. The F.D.A. called the disease 'devastating and deadly' and said the new treatment fills an 'unmet need'".

Grady (2017) also wrote: "There are drawbacks to the approach. Because Kymriah can have life-threatening side effects, including dangerous drops in blood pressure, the F.D.A. is requiring that hospitals and doctors be specially trained and certified to administer it, and that they stock a certain drug needed to quell severe reactions. Kymriah, which will be given to patients just once and must be made individually for each, will cost $475,000. Novartis said that if a patient does not respond within the first month after treatment, there will be no charge. The company also said it would provide financial help to families who were uninsured or underinsured."

In 2016, NIH had about 400 cooperative research agreements with companies. Richtel and Pollack (2016) wrote: "Kite executives and national health officials characterize their partnership as a model arrangement in a system established by Congress three decades ago. The system has given birth to the cancer drug Taxol, the AIDS drug Prezista, two cervical cancer vaccines and a widely used test for H.I.V. infection, among other innovations... The reliance of private companies on government-funded research goes well beyond obvious cases like Kite. In many instances, companies work with universities or medical centers that, in turn, have been funded from the $32 billion annual budget of the National Institutes of Health. Kite's two main competitors, Novartis and Juno Therapeutics, for instance, derived similar immunotherapy treatments largely from academic institutions, developed at least in part with government funding. Novartis has a relationship with the University of Pennsylvania, and Juno with the Memorial Sloan Kettering Cancer Center, the Fred Hutchinson Cancer Research Center and Seattle Children's Hospital. 'For the most important drugs you'll see some public-sector involvement,' said Bhaven Sampat, an associate professor of health policy and management at Columbia University."

Bhaven Sampat had studied recent drug development in the U.S. and found that 48% of new drugs had, at least indirectly, some federally funded research. For medically important drugs, Sampat found that 17% had direct government support (Richtel & Pollack, 2016). Richtel and Pollack (2016) also wrote: "Partnerships and licensing deals were encouraged by the 1980 Bayh-Dole and Stevenson-Wydler Acts, and the 1986 Federal Technology Transfer Act. The laws are credited with jump-starting the biotechnology industry."

National Institutes of Health CRADA

NIH posted: "Cooperative Research and Development Agreements (CRADAs) provide an exciting opportunity for NIH investigators to join with their colleagues from industry and academia in the joint pursuit of common research goals. Government scientists can leverage their own research resources, as well as serve the larger mission of NIH, to facilitate the development and commercialization of health-care pharmaceuticals and products. Companies also can leverage their own R&D efforts while collaborating in state-of-the-art NIH research. The purpose of a CRADA is to make Government facilities, intellectual property, and expertise available for collaborative interactions to further the development of scientific and technological knowledge into useful, marketable products." (NIH, 2017)

NIH also posted: "The Institutes of NIH each conduct their own CRADAs; and each institute has a technology development coordinator to assist in formulating cooperative research agreements, who should be consulted at an early stage of collaboration by the company and the NIH investigator to assist in identifying and developing the proper documents and obtaining the required approvals." (NIH, 2017)

And in another post from NIH: "The intent of Congress in establishing CRADAs was to promote national technological competitiveness and the rapid transfer of the fruits of innovation to the marketplace. CRADA research and development at the NIH should be directed to the development of biological and behavioral technology, products, and processes by transferring relevant knowledge acquired from NIH research efforts to state and local governments, universities, and the private sector." (NIH, 2017)

And NIH posted: "Advances in biomedical and behavioral research depend on a continuum of research efforts, from those aimed at discovering new knowledge, to those aimed at expanding existing knowledge, to those aimed at developing new procedures and products. There is not always a clear distinction among these research activities to allow a concise definition of which activities are appropriate for a CRADA. All CRADAs must be consonant with the primary biomedical research mission of the PHS, and the specific laboratory involved, ensuring that no aspect of that mission is compromised. For example, a proposed CRADA would not be appropriate if the fundamental mission of the NIH is compromised by creating, either explicitly or indirectly, more than minimal constraints on research freedom and communication." (NIH, 2017)

NIH added: "Although there is no restriction on the topic of research appropriate for a CRADA, all CRADA research projects must be highly focused and delineated, and each proposed CRADA must be carefully assessed for its overall research objectives. In considering a proposed CRADA, HHS operating components will determine if the objectives of a proposed collaboration warrant the establishment of a CRADA, or if its goals are more appropriately met through a procurement contract, material transfer agreement, cooperative agreement or other contractual mechanism. Also, the proposed collaborator's scientific and business capabilities will be assessed." (NIH, 2017)

Summary

The U.S. CRADA program shifts between its government funding agencies and industrial firms assisted industry in order to gain access to and participate in cutting-edge science, as funded and performed by government science agencies. This CRADA research cooperation is sketched in Figure 5.1.

Since government is primarily responsible for funding scientific progress and industry is responsible for funding

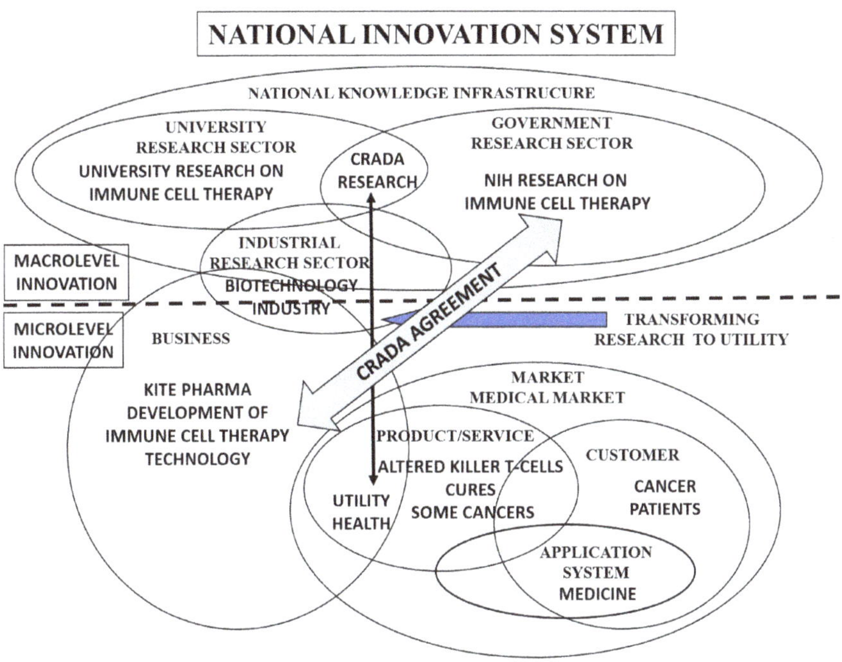

Figure 5.1. National innovation system depiction of CRADA research between NIH and Kite Pharma.

technological progress, a research cooperation program such as CRADA facilitates the rapid transfer of scientific knowledge into technological knowledge. And as we saw earlier, in the innovation process, utility to society from scientific progress occurs through the translation of knowledge from science into technology. The biotech firm of Kite Pharma cooperated with NIH researchers to transfer the science of altered T-cells into a technology of cell therapy to target cancer cells. (Preliminary results at this time in 2017 indicate that altered T-Cell therapy can completely kill some cancers in treated patients.)

CHAPTER 6

R & D Cooperation: University & Industry

Introduction

In this chapter, we examine the cooperation between research and development departments in universities and industry to facilitate innovation through the development of new products or services. Industry is the research sector which produces the commercial products/services in a private economy. Now we review the procedure by which industry transforms research into commercial utility. Industry calls this procedure 'Research & Development' (R&D). In industrial R&D, procedural steps are called 'basic research', 'applied research', 'technology development', and 'commercialization'. R&D stages in the commercialization of science and their relative costs are illustrated in Figure 6.1.

The dark curve indicates the cost in innovation through each stage of R&D; as a product goes through successive stages in the process of product development, the cost increases. Technology 'research' begins with an invention that manipulates nature. Next, technology 'development' improves the performance and safety of the technology for application in a product/service. 'Research' reduces the technical risk of new technology; and 'development' reduces the commercial risk. The rapid increase in cost during the commercialization stage is due to investment needed to build the production process (factory) for the radically new product.

The critical point at which commercial risk (can the business make money from the research) begins to dominate (as

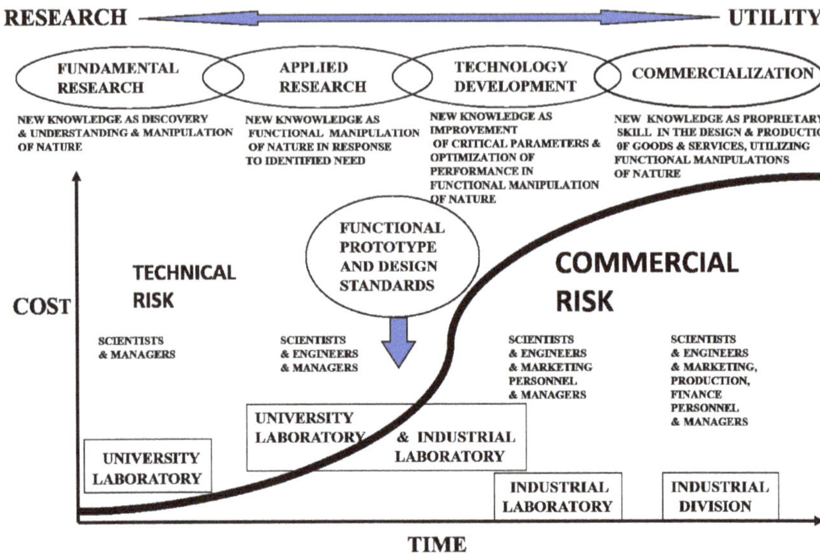

Figure 6.1. The radical new-product realization process.

opposed to the technical risk of will the technology work) is when a functional prototype (or industrial design standard) is reached. In the functional prototype stage, engineering research and engineering development work begins. Following the successful engineering design of a new product or service, commercialisation begins (with research now requiring the participation of not only scientists and engineers, but also marketing personnel and managers to target the commercial sector). In the engineering and commercialisation stages, the cost of research, development, design and production begin to escalate from previous research costs by orders or magnitude — ten times or more in cost).

In the commercialisation stage of innovation, cost continues to rise rapidly as the design of a new product (or service or process) is developed and technical problems must be resolved to make the new technology work in the design. Here, engineering and marketing personnel dominate in this design phase with scientists playing a minor role and only being called upon to solve technical problems encountered in the application of the new technology to the new product (or service or process).

Most of the money a business can lose upon the commercial failure of a new product may be lost here in the commercialisation stage, when the new product fails to deliver a performance or price advantageous to a customer against competitive products. This is the commercial risk in innovation. Therefore industry/university research cooperation, which focuses on reducing the risk of the commercialization stage, contributes greatly to mitigating the cost of industrial innovation.

All R&D stages in innovation are complicated. The innovation process that transforms science into commercially successful utility includes difficult and expensive sequential procedures — from science to technology, to engineering, to commercialization. The cold hard fact of innovation is that it has historically taken from two to ten years to do this, and involved millions of dollars in research and development funds.

German Fraunhofer Society

The Fraunhofer Society (Fraunhofer Gesellshaft, FhG) was founded in 1949 in West Germany to perform applied research for industry. In 1997, the US National Academy of Engineering studied the German R&D system, viewing it as one of the most successful forms of university-industry research cooperation models in reducing the costs of innovation: "The FhG is the major German non-profit organization in the area of applied research, running 46 institutes... The FhG employs 7800 people, of whom 2600 are scientists or engineers." (Abramson *et al*, 1997)

The FhG consists of a society and its institutes. The institutes perform research, and the society governs the institutes. The Institutes form a Member's Assembly of the Society, which elects a Senate for the society. The Senate is responsible for policy, which the Board of Directors administers. The Board of

Directors is composed of the president of the Fraunhofer Society and two full-time directors.

In 1997, there had been forty-six institutes in Germany, six in defence and forty in commercial sectors, with an average staff strength of 170, including part-time employees and students. Research areas of the institutes included: materials, production processes, information & communication, microelectronics, sensors, energy & environment, and services. Funds came from base funding by the society, project research for the government, and contract research for industry. Research covered the development of key technologies, the application of technology in products, the development of prototype products and processes, and the pilot production installations.

The Academy of Engineering reported: "The success of the Fraunhofer model... includes the decentralized management and substantial autonomy of the institutes, which are prerequisites for flexible adaption to the needs of the research market... and direct linkage of the level of institutional funding to success in contract research, which is a major incentive for market orientation and entrepreneurial behaviour... In the German debate on research policy, success with industrial contracts is often seen as the defining feature of the Fraunhofer model, and close linkage to science is overlooked. Both elements, however are important..." (Abramson *et al*, 1997)

In 2010, Antonie J. Jetter (2011) described an institute at Aachen University as the Fraunhofer Institute for Production Technology (ITP), and the Laboratory for Machine Tools and Production Engineering (WZL). These two facilities were located in the Faculty of Mechanical Engineering at Aachen University (which consisted of 50 professors and 6700 students and 2200 employees). They were set up to cooperate in projects in manufacturing research and development. They both specialized in production, and were directed by five chaired professors.

- Professor Kloche directed research in process technology, manufacturing technology, and gearing technology;

- Professor Brecher directed research in production machines, and machine tools;
- Professor Schmitt directed research in metrology and quality management;
- Professor Schuh directed research in technology management and purchasing, and production engineering;
- Professor Sharon directed research in manufacturing innovation.

The budget in 2011 for both ITP and WZL was 32 million euros, of which 40% was from public funds and 37% from industrial projects, with 23% base funding from the Fraunhofer Society and Aachen University.

ITP focused on research on process technology, production machines, metrology (measurements) and quality management, and technology management and purchasing. WZL focused on research on manufacturing technology, machine tools, metrology, and production engineering;

- Process technology research focused upon machining and material removal processes, laser materials processing and forming processes.
- Production and machine tool research focused upon the design of machines, control and testing.
- Gearing technology research focused upon gear manufacturing and gear calculation.
- Management research focused upon business development techniques, technology management techniques, innovation management techniques, and production and quality management techniques.

In industry, applied research must be embedded in a product (or process or service) development project. It is the sales of products and services that earn industrial income (not the sales of technology). The Fraunhofer Institutes helped industrial firms to translate technological progress into product development. This translation required applied research to be carried out so that the technology could be adapted to a product requirement, which led to the design of a new product. The commercial advantage of a new high-tech product/service was in providing a competitive advantage of higher quality and/or lower costs when compared to existing products. A Fraunhofer Institute focussed on product and production development, and in doing so had provided the test-bed for the introduction of new prototype production processes into the industry.

In 2010, Germany had 80 Fraunhofer Institutes, with 18,000 technically qualified employees, and an annual budget of about 1.65 billion euros. This budget came from about one third base-funding from the government's Fraunhofer Society, one third from project contracts, and one third from industrial contracts. The project funding came from German federal and state sources (e.g. the Ministry of Science and Technology, the Ministry of Economics and Employment, etc.). Projects arise from research and development (R&D) proposals, which an industrial and university partner define and write. The industrial and university partner then apply to a government agency for their funding. (Jetter, 2011)

The Fraunhofer Society and Small-Medium Enterprises (SME)

Antonie J. Jetter wrote: "Fraunhofer is an extremely valuable resource for industry, especially for mid-size companies. Fraunhofer shapes industries through services and its influence on government research programs. Fraunhofer tends to make strong industries stronger." (Jetter, 2011) This connection between German universities and mid-sized companies has been one of many important factors in turning the German economy around so soon after World War II.

German technical universities assisted in the growth of the economic strength of German SMEs. The innovation of next-generation technologies usually involves large companies cooperating with universities. Large companies have corporate research laboratories, which can employ scientists capable of basic research who interact with the university scientists. However, the cooperation between universities and SMEs requires a different innovation model. The Fraunhofer Society and German universities connect with German SMEs. The success of the German SME sector has been viewed as a kind of model for small enterprises in any nation — the 'German Mittelstand model', as Meyer-Stamer and Walring (2000) called it.

Except for Germany, the general global pattern between SMEs and large global companies in most nations, has been this: global enterprises create most of the national wealth while SMEs provide most of the national employment. This is one of the paradoxes of present global civilization: wealth versus employment, big versus small businesses. For example, there was a Euro-crisis in Europe in 2010–2013. In most European nations, governments overspent, generating large national debts, while global corporations paid little taxes and, due to a policy of government austerity, unemployment rose in some nations to depression levels. The exception was Germany, which did not suffer a Euro-crisis and had a positive trade balance and low unemployment. *One distinguishing economic feature of this successful German economy was that German SMEs were generating 50% of GDP and 60% of the employment.*

Even in Asia, the SMEs have also encountered the global economic pattern: SMEs are the principle source of employment but not Gross Domestic Product (GDP). For example in South Korea, large Korean global companies operate in chaebol form (large family-owned conglomerates) and provide the majority of GDP revenue, but only a minority of jobs. Korean SMEs generated a minority of the GDP revenue, but provided the majority of jobs. In 2012, chaebols generated 75% of Korea's GDP and provided only 25% of employment. Conversely, South Korean SMEs generated only 25% of GDP but provided 75% of employment. *Wealth versus employment is a serious problem in the modern global world.*

But in Germany, Mittelstand firms (SMEs) created 51.8% of the GDP, 79.2% of German employment, and hosted 83.2% of apprenticeships in the German educational/training system for new workers (Robaschik, 2013). The German term 'stand' refers to the status of guild workers in medieval Germany, who handcrafted the goods produced for trade. The term 'mitte' means 'middle', so that 'mittelstand' refers to a feudal middle estate between the military aristocracy and the peasants of the feudal German-speaking territories. Guilds were a middle class between the rulers and the agricultural ruled of European feudal society.

In the late 1800s, industrialization appeared in Germany, yet small craft shops still existed up till the First World War. In the nineteenth century, large German firms had been growing and were established in steel (Krupp), electricity (Siemans), chemicals (IG Farben) and pharmaceuticals (Bayer). These continued after the First World War, but the German economy was in turmoil during the 1920s and into the 1930s. This was partly due to the penalties and repatriations imposed by the victorious allies of the war, and also to the 1930s global depression. Nazis seized power and began World War II. After losing the war, a devastated Germany was divided into two parts, West and East Germany. In West Germany, small family-owned businesses played a major role in the rapid and successful rebuilding of the industrial economy. The West German government created a national partnership by helping the SMEs to rebuild Germany. A new 'Mittelstand SME culture' was developed in West Germany.

In that commercial culture, German Mittelstand companies are focused on principle product lines, and are not run as business conglomerates. They concentrate on technological innovation to improve the quality and lower the costs of their specialized products/services. They help in education and utilise German labour, making them one of the most highly trained work forces in the world. Using the German university system, they also acquire and develop new technology. They finance themselves conservatively and employ professional management to assist in the running of their businesses.

Mittelstand firms are particularly export-oriented, aiming to dominate high-quality, technically advanced niche markets in the global economy. In the European Union (EU) manufacturing industries in 2009, German SMEs captured 34% of the EU machinery and equipment business, 28% of the metal products business, and 27% of the rubber & plastic goods business. In 2009, there were 340,000 Mittelstand SMEs exporting goods and services to global markets, from industries including manufacturing, retail and wholesale trade, to scientific and technical services. To achieve the technical competitiveness necessary for exporting, the German SME Mittelstand business culture was active in Research and Development (R&D). In 2010, Mittelstand businesses employed 70,000 R&D personnel and spent 9% of their budgets on R&D. (Robaschik, 2013)

The German Federal Ministry of Economics and Technology cited the following reasons for the success of the German Mittelstand:

- "Almost all of Germany's SMEs are family-owned, and many are managed by their owner. Their business policies tend to be especially long-term.
- "The German Mittelstand companies are some of the most innovative in Europe; 54% launched an innovation onto the market in the years 2008–2009.
- "The German Mittelstand relies on sound financing models — chiefly equity and bank loans. The high equity ratio and a cautious approach to expansion enable the companies to undertake medium-term and long-term investments, even in times of crisis.
- "The German Mittelstand employs 83.2% of trainees, the much sought-after skilled workers of tomorrow.
- "The government supports the German Mittelstand on key issues like: investment in R&D, the skills shortage, foreign trade and investment, financing needs, company start-ups and company hand-overs."

(Federal Ministry of Economics and Technology, 2013)

Innovation Intermediary: Modeling a Mittelstand Innovation Structure

Antonie J. Jetter and Songphon Munkongsujarit (2013) discussed the concept of an 'intermediary organization' to facilitate the technology transfer from

(1) Science and technology at a national level, down into
(2) The technology acquisition and development at a firm level — this operating as an *Innovation Intermediary*.

They wrote: "There are a number of different terms in the literature that are used to refer to and to describe innovation intermediary. These terms include: intermediaries, technology brokers, knowledge brokers, innovation brokers, bricoleurs, boundary organizations, matchmakers, and open innovation accelerators. One of the most commonly used definitions of an innovation intermediary is from J. Howells [Howells, 2006] who defines an innovation intermediary as 'an organization or body that acts as agent or broker in any aspect of the innovation process between two or more parties'."

The Fraunhofer Institutes in German universities have played an important 'innovation intermediary' role in connecting university research to industrial development. Each Fraunhofer Institute is an applied research institute at a German university, directed by a chaired German professor who may also direct a basic research institute. The Fraunhofer Society funds the applied research institutes, while the Max Planck Society funds the basic research institutes. Both Societies are research-funding agencies of the German government. Applied research focuses upon advancing technology, and basic research concentrates on advancing science. The Max Planck and Fraunhofer agencies are very important to the science & technology policy of the German government. Together, they provide a pathway that connects the basic research of science to the applied research of technology.

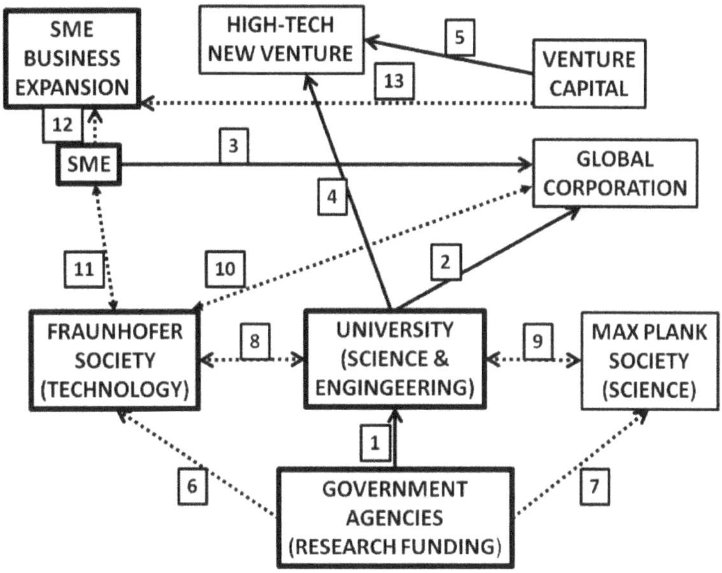

Figure 6.2. The innovation intermediary model in Germany.

The innovation intermediaries of the Max Planck and Fraunhofer Societies have played an important role in providing the German Mittlestand a competitive edge — SME success in helping Germany create both employment and exports. In Figure 4, we use the German examples of the Fraunhofer Society and the Max Planck Society as a model to describe an innovation intermediary infrastructure, with 'innovation intermediaries' between university and industry.

The solid lines express the research connections between government, university, and industry in the German innovation system.

(1) Government agencies provide research funding to universities.
(2) University research advances knowledge in science and in basic engineering and provides both new knowledge and technical graduates to large, global corporations.

(3) SMEs provide supplies and parts to global corporations, as part of industrial value chains. One competitive challenge for a supplier SME is to acquire new technical knowledge to improve technologies for their products.
(4) Universities, in addition, sometimes also contribute to high-tech new ventures when graduates and/or professors begin a company based upon university research inventions.
(5) Venture capital can provide the financing to grow these new businesses.

The dotted lines show the additional research connections introduced by the innovation intermediary of the Fraunhofer institutes. In the German innovation system, the government funds two societies:

(6) The Fraunhofer Society, which funds research progress in technology, and
(7) The Max Planck Society, which funds research progress in science.
(8) The Fraunhofer Society funds Fraunhofer Institutes in universities for engineering research.
(9) The Max Planck Society funds Max Planck Institutes in universities for science research.
(10) The university-based Fraunhofer Institutes provide engineering research for technology and product development to global corporations and
(11) The university-based Fraunhofer Institutes also provide engineering research for technology and product development to SMEs.

This kind of engineering research in Fraunhofer Institutes differs from the science research in Max Planck Institutes. Engineering research can directly produce commercially-designed products and services, and also commercial new production processes and tools. These can be contracted for and provided to both global corporation or SMEs.

In the Max Planck Institutes, science research produces new knowledge published in scientific papers and/or patents, but does not directly produce new products, services, or industrial processes.

A direct dotted-line connection between SMEs and university research is not indicated in the model because university research is usually too far from direct implementation in products or services or production. Global corporations have both the research & development capability and the funds to complete the development of products based on research done. SMEs seldom have such capabilities or access to such funds. This is why Fraunhofer Institutes are very important as innovation intermediaries between SMEs and universities. In the German Mittelstand model, a Fraunhofer and an SME can jointly apply for government funding in a research project. Government funds then pay for 100% of the university effort and 50% of the SME effort. This is a strong financial incentive for SMEs to use Fraunhofer intermediary assistance. An SME can gain the full intellectual value of a new product/process/service, while paying only half its costs of development and no portion of the university's costs. Moreover, university students who helped in developing the new product may be available to be hired by the SME as part of the acquisition of a new technology.

This connection between SMEs and innovation — through the important intermediating contribution of a university engineering Fraunhofer Research Institute — is the technical key to the great success of the German Mittelstand innovation culture.

Summary

In national innovation, businesses earn revenues directly from the sale of products and services, which are derived immediately from engineering research, but not immediately from science. This is why Fraunhofer Institutes (acting as innovation intermediaries) have been very important to German national economic success. Fraunhofer Institutes directly assist German businesses, large and small, with access to science, so that they can research and develop (R&D) commercially successful new products, processes, or services — all of which directly produce sales to generate business revenues.

Hence, science indirectly benefits industry; while engineering directly benefits industry. Yet engineering research which is not based on new science often does not create a competitive advantage. This is why both Fraunhofer and Max Planck Institutes at German universities have been the two pillars of the modern German economy as innovation intermediaries — *technological progress (embedded in engineered products/processes/services) based upon scientific progress.*

The challenge of adapting a 'Mittelstand' innovation model from its German culture to other national cultures is this. How can a nation build and adopt both the 'Fraunhofer' and 'Max Planck' approach to innovation intermediaries into their national innovation system?

CHAPTER 7

Scientific Research Techniques

Introduction

When we discuss the science & technology policy of cooperative innovation between government, industry, and university, the emphasis is on facilitating joint progress in both science and technology. Central to this progress is an approach to inquiry known as the 'scientific method'. What is science? What is scientific research? We will review research techniques central to the scientific method in this chapter.

Science, with its unique epistemological method, was created in Europe in the 1600s. Knowledge existed in the world before then, but not science. The 'methodology of science' was created as a result of several European researchers, beginning with Copernicus in the late 1500s and culminating in the late 1600s with Isaac Newton's grand synthesis of classical mechanics. Newton's paradigm provided the scientific approach to modern physics, depicting material nature as 'mechanisms'. During the 1700s, scientific knowledge in physics, chemistry and mathematics improved by leaps and bounds while enormous scientific progress was made in biology in the 1800s and 1900s.

The Origin of Science

There were many important intellectual precursors to science. For example, alchemy was a precursor to the modern scientific discipline of chemistry, but it was not science; it was a confusion of practices and ungrounded theory. In medieval Europe, the universe was thought to be basically composed of air, earth, fire, and water — alchemy. But now in modern Europe, the universe is thought to be built fundamentally out of energy and mass, atoms and molecules, fields and particles — which are the foundation of chemistry and physics.

Another example would be the modern science of mathematics, which had its important historical roots in Egyptian, Greek and Arab geometry and algebra. But algebra and geometry were not integrated until 1619, when Renes Descartes created the modern mathematical topic of analytic geometry. Nor was the modern topic of calculus created until Newton added the ideas of a differential calculus of infinitesimals to analytic geometry in 1693. (At about the same time but independently, Leibnitz contributed the ideas of integral calculus.) The modern discipline of mathematics grew in the 1700s as mathematicians built upon a modern analytical foundation: geometry, algebra, calculus, vectors, set theory, Hilbert spaces, etc.

The essential difference between civilizations before and after the origin of science is their conception of nature. Before, mankind perceived nature as merely a manifestation of a supernatural being — the supernatural and unobservable — belonging to the world of religion. Afterwards, civilization viewed nature as something that can be observed and studied. Nature is thought about, described, and explained through experiments and theory and scientific paradigms. No longer do we live in a world of superstition and magic. We live in a modern world of science and technology — without magic.

Modern science is both method and paradigms. Newton synthesized scientific method as theory-construction-and-modelling-upon-experimental-data, and also created the first scientific paradigm — *Mechanism*. Briefly, Newton's synthesis of the scientific paradigm of mechanism occurred from a sequence of researches by six particular individuals: Copernicus, Brahe, Kepler, Galileo, Descartes, Newton.

These individuals are particularly important as they mark the first time in history where all the component ideas of scientific method came together and operated fully as *empirically-grounded theory*:

(1) A scientific model that could be verified by observation (Copernicus),
(2) Precise instrumental observations to verify the model (Brahe),
(3) Theoretical analysis of experimental data (Kepler),
(4) Scientific laws generalized from an experiment (Galileo),
(5) Mathematics to quantitatively express theoretical ideas (Descartes and Newton),
(6) Theoretical derivation of an experimentally verifiable model (Newton).

Figure 7.1 sketches the views of the different participants in observing the nature of the sun and earth — resulting in the construction of the scientific theory of the solar system.

(1) Copernicus imagined that if the earth orbits around the sun, then the calculations for an astronomical almanac could be simplified from the theory that the sun orbits the earth. Via

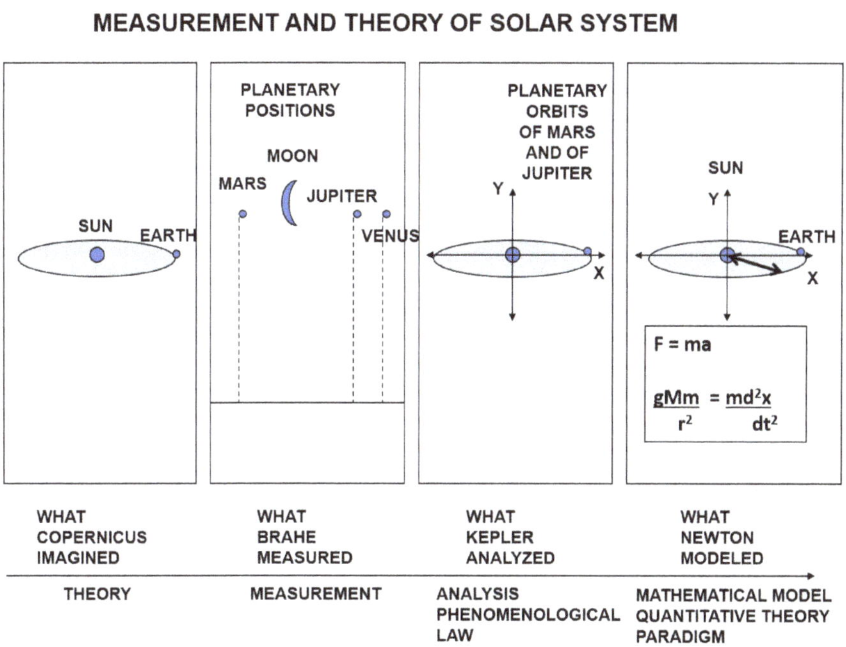

Figure 7.1. Conceptualization physical theory of the solar system.

deductive interference, he proposed a theory of a sun-centric planetary system.

(2) Brahe decided to test Copernicus's theory empirically, using more accurate instruments to find evidence to support it. He improved on astronomical measurements of the appearance of the planets Mars, Jupiter and Venus throughout the year by measuring their planetary positions. Brahe hired Kepler to try to fit the planetary data to circular orbits around the Sun;

(3) Kepler found that the data fit an elliptical orbit instead of a circular orbit. Applying inductive reasoning, he analysed the data and eventually proposed a phenomenological law (Kepler's law) about planetary motion.

(4) In a similar way, Newton applied deductive inference to develop new mathematics (differential calculus), model the solar system and create theories of Newtonian mechanics. He also used inductive inference to propose a theoretical law for gravitation (varying as the inverse square of the distance).

Scientific Method Techniques

Scientific method is fundamental to all research funded by governments or industries. Key ideas in the scientific method are 'empiricism' and 'theory'. Empiricism relies on observations (seeing nature) or experiments (seeing how nature changes under experimental stimuli) for data which we can use to advance our knowledge of nature. Theory is the generalization of the knowledge derived from the observations/experiments conducted such that said knowledge applies to any situation in the area of nature that has been investigated — past, present, or future. For example, the Newtonian theory of the solar system (i.e. that the sun's central gravitation field holds the planets in orbits) predicts that the sun will rise daily, for both today and tomorrow, as it did in the past.

The theory gives the assurance that we understand what exists in nature and how it works. Basing theory on observation

EMPIRICALLY-GROUNDED THEORY

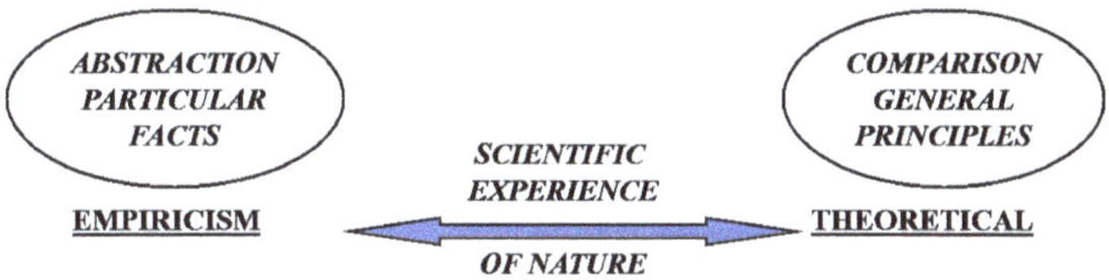

EXPERIMENT	INSTRUMENT
MEASUREMENT	NATURAL OBJECT

ANALYIS	LAW
MODEL	THEORY

Figure 7.2. Empirical and theoretical research techniques in scientific method.

or experimentation allows one to validate if a theory is really true; i.e. whether theory accurately depicts reality. *Theory constructed upon and validated by natural observations/experiments is called Empirically-Grounded Theory, and this is the essence of the Scientific Method.*

There are several research techniques used in constructing empirically-grounded theory and Figure 7.2 lists the different methodological techniques used in the empirical and theoretical aspects of scientific research.

- Experiment is the controlled observation of nature — experiencing nature through the human senses aided by scientific instruments.
- Instrument is a device, which provides a means of extending the range and sensitivity of human sensing of nature.
- Measurement is an instrumental technique in observing nature that results in quantitative data about the phenomenal thing.
- Natural object is the concept identifying a field of observable phenomenon as a class of observable objects (such as atoms, planets).
- Analysis is inferring a mathematical pattern in the quantitative data from a set of experiments.

- Phenomenological Law is a generalization of relationships observed between natural objects.
- Model is a symbolic simulation of the processes of a phenomenal object.
- Theory is a symbolic description and explanation of a phenomenal field of objects.

To see these research techniques in an actual case, we will review the scientific history of the discovery and modelling of DNA, which occurred over a century from the 1850s to the 1950s.

Case History: A Century of Research Culminating in the Modelling of DNA

The science of molecular biology begins with the modelling of DNA, which occurred as a sequence of research projects over a hundred years. The scientific issue which guided that research sequence was the question: How is life reproduced?

In the scientific method, the first issue in any research project is to formulate a question about nature. Questions about the basics of nature (questions of science) consist of *inquiries* into nature. What things exist in nature? How do they interact?

In molecular biology, the 'natural object' studied was 'life'. A generational interaction between the phenomenal things-of-life (organisms) is reproduction, or 'heredity'. The scientific answer to the question of heredity required many events-of-scientific-research, many research projects, which were performed over a long period of time:

1. Investigating the structure of the cell.
2. Isolation and chemical analysis of the cell's nucleus, DNA.
3. Establishing the principles of heredity.
4. Discovering the function of DNA in reproduction.
5. Discovering the molecular structure of DNA.

1. The Structure of the Cell

The history of DNA began in the 1800s, with the investigation of the structure of biological cells (Portugal and Cohen, 1977). Scientists used a new invention, the microscope, to look at bacteria and cells. Cells were seen as the constituent modules of living beings. The technique of using the microscope (technology) to observe nature led to the discovery of the cellular structure of life.

The microscope itself arose from optical technology in the eighteenth century and became used as an instrument in science. All instruments in science are technologies. The technology of the microscope was invented in the 1600s; Hans Lippershey, Hans Jenssen, and Zacharias Jenssen (Dutch eyeglass makers) are given historical credit for the optical lens development of the first telescopes, which later became inverted into microscopes.

In 1838, Christian Ehrenberg (1795–1876) was the first to observe the division of the nucleus during cell reproduction (Portugal and Cohen, 1977). Ehrenberg was born in Germany and completed his doctoral work at the University of Leipzig in 1818. From 1820–1825, he collected specimens of plants and animals on an expedition to the Middle East. He was appointed professor of medicine in 1827. He accompanied Wilhelm von Humboldt on an expedition to the Chinese frontier of Eastern Russia, and refocused on studies of microscopic organisms on his return.

In 1842, Karl von Nageli (1817–1891) observed the rod-like chromosomes within the nucleus of plant cells. Von Nageli was born in Switzerland and studied in the Universities of Zurich and Jena, focusing on the microscopic study of plants. He was a professor at the Universities of Zurich, Freiburg and Munich.

By the middle of the nineteenth century, biologists had seen that the key to the biological reproduction of life involved chromosomes, which divided into pairs when the nucleus of the cell split in cell reproduction. The first significant scientific event toward the discovery of DNA had been the description of the structure of a biological cell, using the microscope to look at bacteria and cells. "Cells" began to be theoretically thought of as the constituent modules of all living beings.

We can now use the case of the scientific discovery and modelling of DNA to understand some of the terms we use in 'research techniques'.

- Experimental observations in biology were extended by the Instrument (microscope) to view events happening on a smaller scale (microscopic).
- This provided experimental observations of units-of-life that were abstracted into the idea of a natural biological 'cell' (Natural Object) and generalized for all organisms into a new underlying form (Theory).
- Known as 'cell theory', Christian Ehrenberg was the first to observe the division of the nucleus when a cell reproduced in 1838, and
- In 1842, von Nageli observed rod-like chromosomes (Natural Object) within the nucleus of plant cells. All this can be seen in Figure 7.3.
- Biologists had seen that the key to the cellular Natural Object in the biological reproduction of life involved chromosomes, which divide into pairs when the nucleus of the cell splits in cell reproduction.
- Further biological Experiments resulted in the formulation of the process (Laws) of cellular reproduction, Figure 7.4.

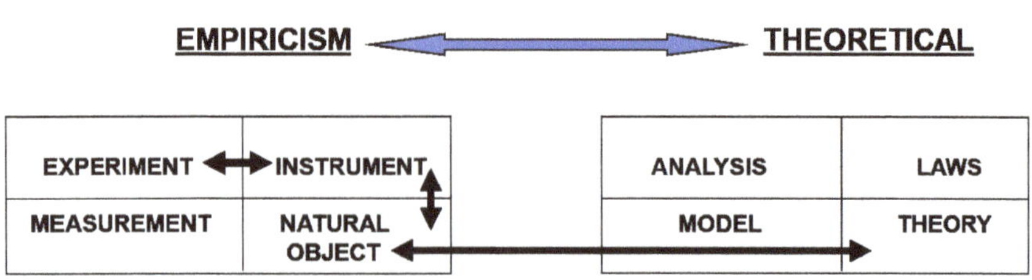

Figure 7.3. Research techniques in conceptualizing 'cell theory'.

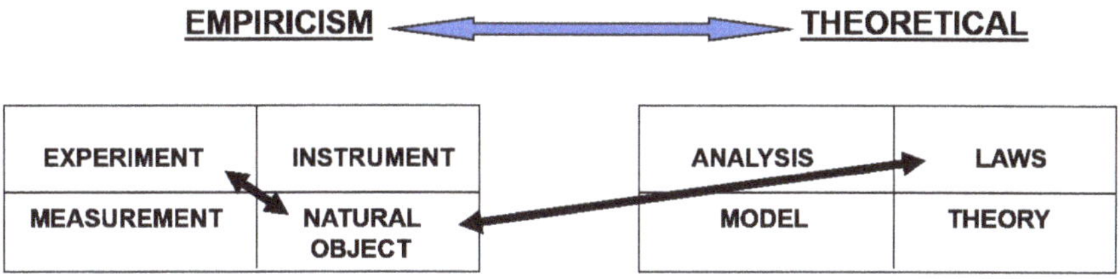

Figure 7.4. Research techniques in discovering the law of cellular reproduction.

2. Discoveries and Chemical Analysis of DNA

We next describe the scientific progress of the chemical analysis of DNA. In 1869, a Swiss chemist Friedrich Miescher (1844–1895) reported the discovery of DNA by precipitating material from the nuclear fraction of cells. He isolated the nucleus of a cell and determined that it chemically contained phosphorus, nitrogen and sulphur. He called the material 'nuclein'. Subsequent studies showed that it was composed of two components, nucleic acid (later called DNA) and protein. After obtaining a medical degree at the University of Basel in 1886, Miescher turned to physiological chemistry and was appointed professor at the same university. He and his students studied the chemistry of nucleic acid, but still did not understand its specific cellular function.

Scientific attraction then turned to investigating the chemical nature of chromosomes. Science is organized into disciplines, but the techniques and knowledge in one discipline may also be used in another discipline. This is one way that the organization of science interacts with the philosophy of science: disciplinary boundaries and the transfer of theory and techniques across disciplines. Chemistry and physics provide theory and techniques (methods) to the scientific discipline of biology.

While Miescher's studies were happening, there were continual improvements in microscope techniques — *scientific method*. The more detail one wishes to observe, the more the means of observation of science needs to be improved. For the microscope, specific chemicals were found that could be used to selectively stain the cell. In the 1860s, Paul Ehrlich discovered that staining cells with the new chemically-derived, different coal-tar colours allowed him to identify the chemical composition of the cell components. As new products of the chemical industry, the new colours are an example of how new products of technology can contribute to science. Progress in technology underlies progress in science. But progress in technology can also provide instrumental bases for progress in science. This is the interaction of *scientific method and scientific application.* Ehrlich (1854–1915) was born in Prussia and obtained a medical degree at the University of Leipzig. He joined the Institute for Infectious Diseases in Berlin in 1886. In 1896, he became director of the Institute of Serum Research and Examination in Berlin and received the 1908 Nobel Prize for Medicine for developing new drugs; amongst them were drugs to treat sleeping sickness and syphilis (called Salvarsan).

From Ehrlich's staining of cells, new techniques for Instrument use improved Experimental Observations at the microscopic level of the Natural Object, better revealing the structure (Model) of the cell (Natural Object), as shown in Figure 7.5.

In 1873, A. Schneider described the relationships between the chromosomes and various stages of cell division. He noted two stages in the process of mitosis. Mitosis is the phenomenon of chromosome division resulting in the division of the cell

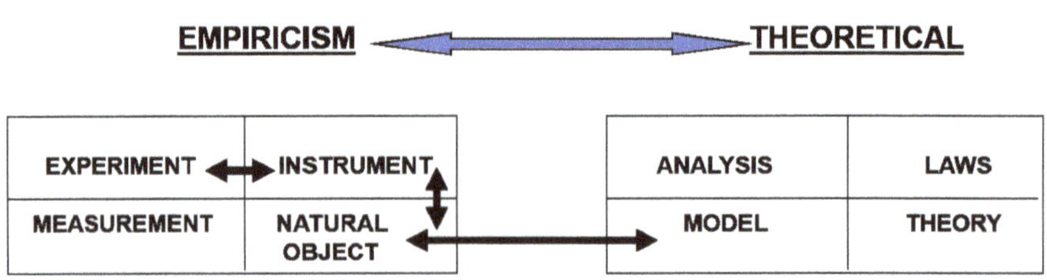

Figure 7.5. Research techniques in revealing structural model of the cell.

nucleus into two daughter nuclei. The cell division process of "mitosis" continued to be described from experiments as a process (a natural process).

In 1879, Walther Flemming (1843–1905) also studied cell division and introduced the term 'chromatin' for the coloured material found within the nucleus after staining. He suggested that chromatin was identical with Miescher's nuclein. Flemming was born in Germany, graduating in medicine at the University of Rostock in 1843. In 1876, he became a professor of anatomy at the University of Kiel. He used aniline dyes to identify the chromatin, investigated the process of cell division and the distribution of the chromosomes to the daughter nuclei, and called the process 'mitosis'.

At this time, studies of nuclear division and the behaviour of chromosomes emphasised the importance of the nucleus. But it was not yet understood how these processes were related to fertilization. In 1875, Oscar Hertwig (1849–1922) demonstrated that fertilization was not the fusion of two cells but the fusion of two nuclei — *scientific method interacting with scientific content — epistemology with ontology*. Hertwig was born in Germany and studied at the University of Jena. In 1888, he became a professor of anatomy at the University of Berlin. He discovered fertilization in sea urchins and recognized the role of the cell nucleus inheritance. He found that fertilization included the penetration of spermatozoon into the egg cell.

Thus, Schneider described the relationships between the chromosomes and various stages of cell division in the two-stage process of mitosis. Hertwig demonstrated that fertilization was not the fusion of two cells but the fusion of two nuclei. Experiments showed another Law in cellular theory (Theory), that cellular reproduction involves the fusion of two nuclei, Figure 7.6.

Meanwhile, the study of nucleic acid components also progressed. In 1879, Albrecht Kossel began publishing literature on nuclein. Over the next few decades, he (along with Miescher) was at the forefront of nuclein research. Kossel (1853–1927) was

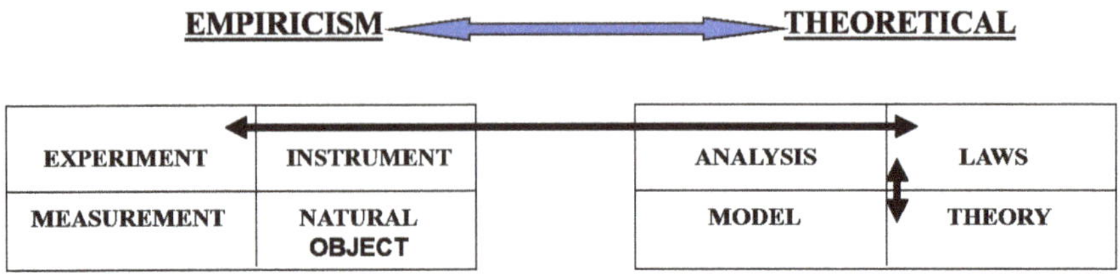

Figure 7.6. Research techniques establishing the law of reproduction in cellular reproduction theory.

born in Rostock as the son of a Prussian consul. He attended the University of Strasbourg to study medicine and graduated later from the University of Rostock in 1878. Kossel worked in physiological chemistry on the chemistry of tissues and cells. He began studying the cell nucleus in 1880 and received a Nobel Prize in Medicine in 1910 for his research in cell biology.

Together, the research studies of Miescher and Kossel, along with Phoebus Levene, finally laid the solid foundation for the determination of the chemistry of nucleic acids. Phoebus Levene (1869–1940) was born in Russia and studied medicine at the Imperial Military Medical Academy, receiving a medical doctorate in 1891. In 1893, he moved with his family to New York in the United States; he was appointed head of the biochemical laboratory at the Rockefeller Institute of Medical Research in 1905. He identified the components of DNA in 1909 and discovered ribose and dexoyribose in 1929.

Back as early as 1914, Herman Emil Fisher had attempted the chemical synthesis of a nucleotide (component of nucleic acid, DNA). Emil Fisher (1852–1919) was born in Germany and earned his doctorate in chemistry at the University of Strasbourg in 1874. In 1881, he was appointed Professor of Chemistry at the University of Erlangen, later moving to the University of Wurzburg and the University of Berlin. He received the Nobel Prize in Chemistry in 1902. He studied purines, sugars and proteins by separating and identifying individual amino acids, the constituents of proteins.

But real progress was not made in the synthesis of nucleotides until 1938 — two decades later. Chemical synthesis of DNA was an important scientific technique necessary to understand the chemical composition of DNA. It took a long time (from 1869–1938) to establish the chemical basis of nucleic acids. *In the organization of science, the timelines for scientific progress are measured in decades, not months or years.*

One of the problems was that DNA and RNA were not distinguished as different molecules until 1938. This is an example of the kinds of problems that scientists often encounter. *Nature is often more complicated than originally thought.* In the cell, DNA records reproductive information but does not directly act as a template to construct proteins. Instead, DNA acts as a template to chemically construct RNA, and the RNA acts as the template to construct proteins. (Why? This is not yet fully known. One conjecture is that perhaps RNA was the more primitive information code, but it was not so robust in evolution and DNA was evolved as more chemically stable, while RNA is still used to make proteins.)

By the end of the 1930s, the true molecular size of DNA had been determined. In 1949, C. E. Carter and W. Cohn found a chemical basis for the differences between RNA and DNA. And finally, by 1950, DNA was known to be a high-molecular-weight polymer with phosphate groups linking deoxyribonucleosides and the 5′ end of one strand aligning with the 3′ end of the other. The sequence of bases in DNA was then still unknown. Thus by 1950, the detailed chemical composition of DNA was finally determined, but not its molecular geometry. Almost 100 years had passed between the discovery of DNA and the determination of its chemical composition. Many Experiments had to be done to finally confirm the chemical Analysis of DNA, Figure 7.7.

3. The Principles of Heredity

From 1900 to 1930, while the chemistry of DNA was being sought, the foundation of modern genetics was also being established.

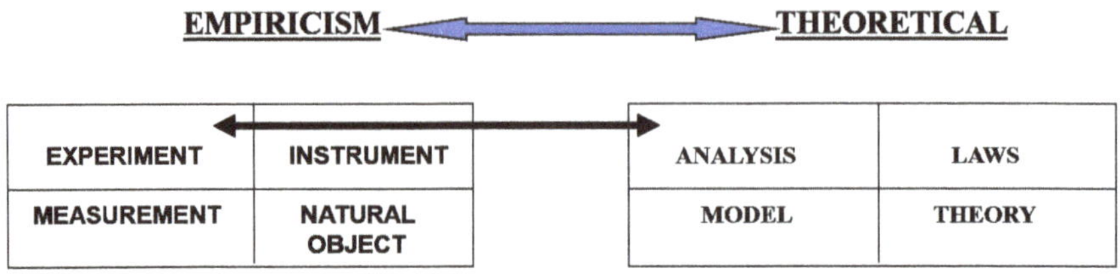

Figure 7.7. Research techniques of experiments to analyze the chemical basis of DNA and RNA.

Understanding the nature of heredity began in the nineteenth century with Darwin's epic work on evolution, and Mendel's pioneering work on genetics. This established a new specialty area in biology — the ontological understanding of heredity. Gregor Mendel (1822–1884) was born in Austrian Silesia (now in the Czech Republic). He studied at the Philosophical Institute in Olomouc, and in 1843 entered the Augustinian Abbey of St. Thomas in Brno. He was christened Johann but took the name Gregor on entering monastic life. He conducted genetic studies in the monastery's garden by cultivating pea plants; his studies showed that one in four plants showed regressive genetic characteristics. He presented his results in a paper at meetings of the Natural History Society of Brunn in Moravia in 1865. He became an abbot in 1868. His pioneering studies have earned him the honour of being called the 'father of modern genetics'.

Modern advances in genetic research began in 1910 with Thomas Hunt Morgan (1866–1956)'s group researching heredity in the fruit fly, *Drosophila Melanogaster*. Morgan demonstrated the validity of Mendel's analysis, and showed that mutations could be induced by x-rays, providing one means for Darwin's evolutionary mechanisms. Morgan introduced physical instrumentation methods into the study of heredity — epistemology interacting with ontology. Thomas Hunt Morgan (1866-1945) was born in the United States and received a doctorate from Johns Hopkins University in 1891. He taught at Bryn Mawr. In 1900, the work of Mendel was rediscovered by

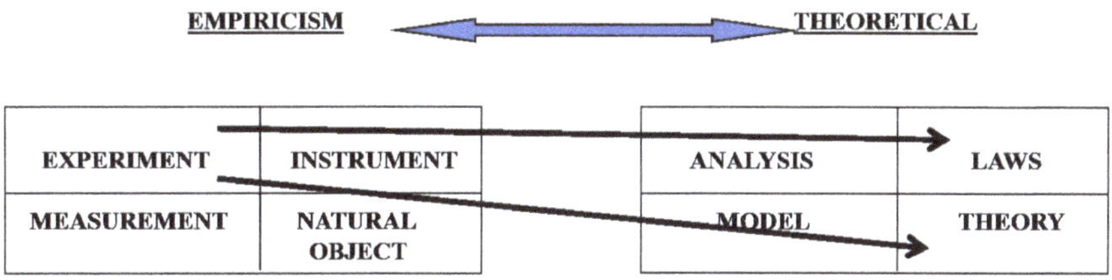

Figure 7.8. Research techniques in establishing the phenomenological law and the theory of inheritance.

scientists. In 1904, Morgan moved to Columbia University, where he began genetic researches on the fruit fly. He demonstrated that genes are carried on chromosomes, which are the mechanical basis for heredity. In 1928, Morgan moved to The California Institute of Technology to establish a new division of biology. He was awarded the Nobel Prize in Physiology of Medicine in 1933.

By 1922, Morgan's group had analysed 2000 genes on the Drosophila fly's four pairs of chromosomes, and had attempted to calculate the size of the gene. Muller showed that ultraviolet light could also induce mutations. *Mendel's pea plant experiments established a phenomenological Law of inheritance of traits*; while Morgan's fruit fly experiments developed the Theory of mutations as a mechanism for Darwin's evolution theory.

4. Function of DNA in Reproduction

While the geneticists were showing the principles of heredity, the mechanism of heredity had still not been demonstrated. Was DNA the transmitter of heredity, and if so, how? Other scientists had been studying the mechanism of the gene, with early work on bacterial reproduction coming from scientists using bacterial cultures.

R. Kraus (1897), J. A. Arkwright (1920), and O. Avery with A. R. Dochez (1917) had demonstrated the secretion of toxins by bacterial cultures. This raised the question of what chemical components in the bacterium were required to produce immunological specificity. The search for the answer to this question revealed a relationship between bacterial infection and the biological activity of DNA.

Also as early as 1892, there had occurred the identification of viruses and their role in disease. In 1911, Peyton Rous discovered that rat tumour extracts contained virus particles capable of transmitting cancer between chickens. In 1933, Max Schlesinger isolated a bacteriophage (virus) that infected a specific kind of bacteria. Next, scientists learned that viruses consisted mainly of protein and DNA. In 1935, W. M. Stanley crystallized the tobacco mosaic virus, which encouraged scientists to further investigate the physical and chemical properties of viruses.

In 1928, Frederick Griffith showed that a mixture of killed infectious bacterial strains with live non-infectious bacteria could create a live infectious strain. He called this a transformation but did not yet understand that it was a transfer of genetic material as DNA. In his experiment, a laboratory mouse was injected with a non-virulent strain of bacteria and the mouse survived. A second mouse was injected with a virulent strain of bacteria that killed the mouse. Then, a heat-killed batch of the virulent strain (bad-but-dead bacteria) was injected into a mouse that survived. This proved that only live, virulent strains of bacteria were lethal. Next, a heat-killed virulent strain of bacteria was mixed with a live, non-virulent (good-and-alive) bacteria, and the mix was injected into a mouse. The mouse died. The only way the mouse could have died was if there was a transfer of genetic material from the heat-killed virulent strain to the live, non-virulent strain, such that the live, non-virulent strain gained virulent characteristics, and was able to infect the mouse. This experiment proved that the virulence characteristic in the bacteria was controlled and transmitted by genes in the bacteria. (Wikipedia, 2018) Frederick Griffith (1879–1941) was born in England and trained as a medical doctor. He tried unsuccessfully to make a vaccine against pneumonia, which was infectiously prevalent after the First World War. He died in the Second World War during an air raid on London.

In 1944, Oswald Avery (1877–1955), along with Colin McLeod and Maclyn McCarty, showed that the Griffith's transformation in bacteria was due to the exchange of DNA between dead and living bacteria. This was the first clear demonstration that DNA carried genetic information. Avery was born in Nova Scotia, Canada but moved to the United States with his family. He received a medical doctoral degree from the College of Physicians and Surgeons in New York in 1904, and joined the Rockefeller Institute in 1923. *This is an example of scientific method, whereby a relationship is established between a mechanism (DNA) and a biological process (infection).*

In 1940, George Beadle (1903–1989) and Edward Tatum (1909–1975) further investigated the mechanisms of gene action, showing that the genes control cellular mechanisms by controlling enzymes, which are needed for chemical synthesis of proteins in the cell. Beadle and Tatum (along with Joshua Lederberg) would later share the 1958 Nobel Prize in Physiology of Medicine.

The stage was now set to understand the structure of DNA, and how its structure could transmit heredity.

While geneticists were examining the function (paradigm) of heredity, other biologists were examining the mechanism (paradigm) of heredity, showing that the mechanism of DNA was the transmitter of the function of heredity. The scientific method of experimentation thus connected the two scientific paradigms, that of the function of heredity and that of the mechanism of heredity, with the result that scientific progress was made. Griffith's experiment demonstrated that DNA passed the infectious function from a dead bacterium to a different live one. Beadle, Tatum and Lederberg demonstrated that genes control cellular processes and that genes could be sexually exchanged between bacteria. *A biological explanation requires*

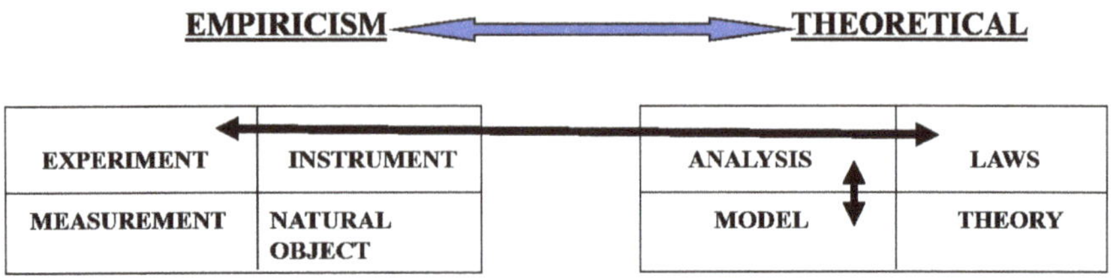

Figure 7.9. Research techniques the mechanistic model in the functional law and the theory of inheritance.

both a mechanism and a function, and a connection between the two. Thus the two kinds of Experiments, one on genetics and another on DNA transmission of gene information, together established that the process (Law) of heredity transmission was through the Model of DNA exchange, as shown in Figure 7.9.

5. Structure of DNA

So far, we have reviewed the long and numerous lines of research necessary to discover the elements of heredity (genes and DNA) and their function (transmission of heredity). Yet before we can understand the scientific applications of this knowledge (technology), one missing component will have to be in place to complete the picture. The mechanism will have to be understood.

This step was achieved by a group of scientists that would later be called the 'phage group'. The discovery of the molecular form of DNA by the scientists in this group would then create the modern scientific specialty of molecular biology.

Historically, bacteriophages (viruses) had been studied in biology since 1917 when they were discovered by Felix d'Herelle. Felix d'Herelle isolated phages (viruses) by

(1) infecting a nutritional medium with bacteria,
(2) infecting the bacteria with phages,
(3) filtering the medium through a porcelain filter which only the small phages could pass through.

Felix d'Herelle (1873–1949) was born in Montreal, Canada. At the age of six, after the death of his father, he and his mother moved to Paris. He began traveling through Europe when he was seventeen and met and married Marie Caire. He moved his new family to Canada where he worked for the Canadian government, studying the fermentation and distillation of maple syrup to schnapps. He next worked as a bacteriologist for Guatemala and Mexico before returning to France. There, he worked in the Pasteur Institute, extracting bacteria harmful to locusts that were plaguing the Mexican production of sisal. During World War I, d'Herelle remained in France, and produced medication for the allied military. In 1917, d'Herelle announced his discovery of 'an invisible antagonistic microbe of the dysentery vacillus' — a phage (virus).

It was not until 1940 when viruses were used to understand genetics. In 1940, Max Delbruck (1906–1981) and Salvador Luria (1912-1991) began collaborating on experiments on different bacteria and bacteriophages, and showed that a given bacteria could only be infected by a specific strain of phage. Delbruck was born in Berlin, Germany. He received a PhD in physics in 1930 from the University of Gottingen, but became interested in biology and moved to the United States in 1937, taking work as an assistant in the Biology Division at Cal Tech under Morgan. He moved to Vanderbilt University to teach physics, but continued genetic research with Salvador Luria at Indiana University. They showed that bacterial resistance to a viral infection was caused by random mutation. For this work, Delbruck and Luria won a Nobel Prize in Physiology of Medicine in 1969 (sharing it with Alfred Hershey). Luria was born in Turin, Italy. He received a medical degree from the University of Turin in 1936. He took classes on radiology at the University of Rome, and read Delbruck's paper. In 1938, he received a fellowship to study in the United States and intended to work with Delbruck. However, Mussolini's Fascist regime banned Jews from academic research fellowships. Hence, Luria went to Paris in 1938. When the German Nazi army invaded France, Luria fled to

Marseilles and immigrated to the United States. In 1940, he visited Enrico Fermi at Columbia University, whom he knew from his time of study at the University of Rome. Fermi helped him obtain a fellowship from the Rockefeller Foundation for work at Columbia.

At Columbia University, Luria met Delbuck and Hershey, and they began collaboration on experiments at Cold Spring Harbor Laboratory on Long Island, New York. This collaboration continued even after Delbruck returned to Vanderbilt at the end of the summer (and eventually won them a Nobel Prize). From 1943 to 1950, Luria taught at the University of Illinois. In 1950, Luria moved to Indiana University, where he discovered that specific enzymes could cut DNA in *E. coli* genes at specific places in the DNA — restriction enzymes. In 1959, he moved to Massachusetts Institute of Technology.

The story behind the discovery of the molecular structure of DNA moves on to James Watson, Luria's first graduate student at the University of Illinois. Born in Chicago, Illinois, in 1928, Watson was educated in the ontological and epistemological approach of the Phage Group. As a graduate student Watson knew that the eventual goal (ontological issue) was to discover the structure of DNA. He received a post-doctoral fellowship to do research in Europe, looking for a way to pursue his research goal. He went first to Copenhagen, then to Italy; there, he attended a scientific meeting where Maurice Wilkins talked about his x-ray diffraction studies on the DNA molecule. In 1946, he had read Erwin Schrödinger's book, *'What is Life?'* reorienting him from ornithology to genetics. He went on to the University of Illinois for graduate work, having read Luria and Delbruck's research on viral mutations (Phage Group). In summer 1948, Watson began research work under Luria, and met Delbruck during Watson's first trip to Cold Spring Harbour Laboratory. Watson's research projects used x-rays to inactivate bacterial viruses (phages), and he obtained his PhD in Zoology from Indiana University in 1950. He graduated with a PhD under Luria in 1951.

While Watson was in Europe, Linus Pauling published a molecular model of a protein in the geometric form of a helix in 1951. Pauling was using x-ray crystallography and molecular model-building. Linus Pauling (1901–1994) was born in Portland, Oregon, USA. In 1917, Pauling attended Oregon Agricultural College (now Oregon State University), and graduated in chemical engineering in 1922. He then went to California Institute of Technology. On obtaining a PhD in physical chemistry and mathematical physics, he then travelled to Europe on a Guggenheim Fellowship to study under the German scientists who had created quantum mechanics: Arnold Sommerfeld at the University of Munich, Niels Bohr in Copenhagen, and Eriwn Schrodinger in Zurich. Pauling became acquainted with the newly growing quantum mechanical theory, which was the key to atomic binding into molecules — the physical basis of chemistry.

In 1927, Pauling became an assistant professor in theoretical chemistry at Cal Tech. In the 1930s, Pauling began publishing work on the nature of the chemical bond and wrote a pioneering text on the subject, which would later win him the 1954 Nobel Prize in Chemistry. Pauling also worked on crystalline structures of biological molecules in the 1930s, using x-ray pictures of protein molecules made by William Astbury in England. In 1948, Pauling modelled haemoglobin, in which atoms are arranged in a helical pattern. In 1951, he modelled amino acids and peptides, proposing both an alpha helix model and beta sheet (flat) model to structure these. This was the 'helix' model, which was to stimulate Watson.

Watson saw himself in a race against Pauling to be the first to model DNA. From Pauling's paper, Watson saw a research strategy for how to model DNA. First, he would need to obtain an x-ray crystallographic picture of the structure of DNA to infer its geometry. Then, he would have to build a chemical model of DNA in that geometric structure. DNA might possibly have a helical geometry, as did Pauling's peptide molecule. Since he had heard Wilkins's report on the x-ray diffraction studies on DNA, Watson decided he had to go England. Watson asked Luria to arrange for his fellowship to allow him to work at Cambridge. *Scientists, as well as managers, perceive competition in their environments. Watson looked at research from a research management perspective — philosophical and organizational. He needed to beat scientific competitors to the goal and needed a proper research facility and team to win the competition.*

Watson had two kinds of intellectual inputs (research techniques) from Pauling's work to put into his research strategy (scientific method). First, Watson had x-ray pictures of the crystalline structure of DNA molecule. Second, Watson had a molecular modelling approach from Pauling, using quantum mechanics and showing a helix form. To use these inputs, he needed to collaborate with both a physicist for the quantum mechanics, and with an X-ray experimentalist for x-ray pictures of DNA. These were Francis Crick and Maurice Wilkins. Crick was in Cambridge, and Wilkins was nearby in London.

Once at the Cavendish Lab in Cambridge University, Watson found a collaborator in Francis Crick. Crick (1916-2004) was then still a graduate student working on his physics degree; he was older than Watson because his graduate studies had been interrupted by service in the Second World War. In 1951, Crick was assisting in the development of the mathematical theory of how a helical model would produce x-ray diffraction pictures. Late in 1951, Crick met Watson and they began working together on Watson's goal to develop a molecular model of DNA. Crick was born in Northampton, England, and obtained a Bachelor of Science degree in physics from University College London in 1937. Crick began graduate work to measure the viscosity of water at high temperatures, but was interrupted when a bomb fell upon the laboratory after World War II had begun. During the war, Crick worked for the British Admiralty Research Laboratory, designing magnetic and acoustic mines. In 1947, Crick turned to biology, working first in Cambridge at the Strangeways Laboratory, and then later in the Cavendish

Laboratory. Influenced by the works of Linus Pauling (chemist) and Erwin Schrodinger (physicist), Crick was interested in researching the electronic bonding of atoms into biological molecules. Crick began working on x-ray crystallography in the Cavendish Lab.

But the x-ray diffraction pictures of DNA were not being taken in the Cavendish Lab at the University of Cambridge. Instead, they were being made at King's College London in the University of London. In 1946, a new biophysics laboratory had been established in King's College in Cambridge, and Wilkins began doing x-ray diffraction studies on proteins. Fortunately, Wilkins was a friend to Crick. With DNA samples from calf thymus, Wilkins was able to produce thin threads of DNA that contained highly ordered arrays of DNA, suitable for the production of x-ray diffraction patterns. Maurice Wilkins (1916–2004) was born in New Zealand, but moved to Birmingham, England when he was six. He studied physics at the University of Cambridge in St. John's College, and received a PhD in physics at the University of Birmingham in 1940. During World War II, he worked to improve radar screens in Britain. He also worked on isotope separation in the Manhattan project in the United States.

But although Wilkins had begun on the X-ray research, while he was away on holiday, the Director of the Biophysics Laboratory, John Randall, hired a new post-doc to continue the research. Her name was Rosalind Franklin. When Wilkins returned from holiday, he assumed that Franklin worked for him. But Franklin assumed she worked for Randall. This difference hampered good working relationships between Wilkins and Franklin — which would have unfortunate professional consequences.

Randall was a research manager and director. Wilkins saw himself as a research manager. But for which manager did Franklin work? This organizational difference was to have profound effects upon the course of the DNA modelling research, and on how scientific rewards were allocated. Rosalind Franklin

(1920-1958) was born in England. In 1938, she entered Newham College in the University of Cambridge, graduating in 1941. She began research in Cambridge under Ronald Norrish (a Nobel Prize chemist in fast chemical reactions). In 1942, she left to do war research for the British Coal Utilization Research Association, studying the porosity of coal. In 1945, she obtained a doctorate degree from the University of Cambridge in the physical chemistry of solid organic colloids, with special reference to coal. At the end of the war, Franklin went to Paris to work under Jacques Mering at the Laboratoire Central des Services Chimiques de l'Etat, learning x-ray diffraction techniques. She then wanted to return to England and accepted a position at King's College London offered by Randall to work on the DNA project.

Earlier in science in the 1930s, X-ray crystallography had been developed as a new scientific instrumental technique. X-rays (high-energy photons) were sent through crystals and formed pictures on photographic negatives of the scattering of these X-rays from the planes of atoms in the crystal's atomic structure. By looking at the X-ray diffraction picture, and calculating, one could describe the crystalline structure in terms of the planar structure of its atoms and the spacing between planes of atoms. Just as the microscope was essential to observe the cell and its structure, x-ray crystallography was essential to observe the structure of DNA.

(As an analogy to photon waves in x-ray diffraction, consider the movement of water waves around the structure of a pier extending out into the ocean. By watching waves come in and under the pier, one can see new waves refracted around the pilings of the pier. Then one can see how the spacing of the pier's pilings produces smaller waves. Then watching these smaller waves, one can calculate backwards to measure the spacing between the pilings. This is how x-ray diffraction calculations work.)

But Franklin's research task was turning out not to be easy. It happened that there were two crystalline forms of DNA, and only one would yield a good picture. Moreover, the crystal in that picture would have to be oriented just right to get a picture that would be interpretable in terms of its structure (seeing the helix of the crystal from just dead on down the axis of the helix). Finally, Franklin did get a good picture of the right form, a picture right down the centre of a clearly helical form. *Creativity in experimental techniques to tease out the secrets of nature is the talent of an outstanding experimental scientist (and very different from the talents of a theoretical scientist).*

Meanwhile, Watson and Crick had begun their modelling research away in Cambridge. Watson brought knowledge of biology and organic chemistry to their collaboration and Crick brought knowledge of physics; both skills were necessary. From Pauling's research showing a helical form of a protein, Watson had conjectured that DNA too had a helical structure. But what kind of x-ray diffraction picture would a helical molecule make? Watson wanted to be prepared to interpret it from an x-ray, and he asked another young expert in diffraction modelling for a tutorial. He was told (if the picture were taken 'head-on' down the axis of the helix) how to see the helix and measure its angle. Watson was thus equipped to interpret an x-ray picture of DNA, if he could get his hands on one. *This is an example of scientific method as connecting observation to theory.*

Crick also needed to know precisely how the chemistry was ordered in a DNA molecule. Fortunately, the one scientist who knew this was Erwin Chargaff. In 1952, Chargaff happened to visit England and in Cambridge lectured about the chemistry of DNA (nucleotide biochemistry). Watson and Crick attended his lecture and realized that they needed to learn more about the nucleotide biochemistry.

Erwin Chargaff (1905–2002) was born in Austria. In 1923, he attended the University of Vienna, receiving a doctorate in chemistry in 1928. He then travelled to the United States as a research fellow and did postdoctoral work in organic chemistry at Yale University. In 1930, he returned to Europe working as a research assistant at the University of Berlin, and then at the

Pasteur Institute in Paris. In 1935, he immigrated into New York as a research associate in the biochemistry department at Columbia University, becoming an assistant professor in 1938 and a full professor in 1952. He published works in nucleic acids, using chromatographic techniques. Chargaff received the Pasteur Medal in 1949 and later the National Medal of Science in 1974. Chargoff had discovered the relative proportions of amino acids that compose DNA. He had chemically determined that the amounts of adenine (A) and thymine (T) are roughly equal (A = 30.9% and T = 29.4% of the bases in DNA), and that the amounts of cytosine (C) and guanine (G) are roughly equal (C = 19.9% and G = 19.9% of the bases in DNA). This is known as 'Chargaff's rules'.

Chargaff's rules were critically important to Watson and Crick's modelling efforts. In February 1953, Watson and Crick began to make the first of their geometric models of the structure of the DNA molecule using cut-outs in white cardboard and paste. The atoms composing the bases, A, T, C and G are all geometrically flat in a ring structure. Watson cut hexagonal rings out of the white cardboard to represent their flat ring structure. Then he paired A and T and paired C and G, looking to see where hydrogen bonds might couple the paired rings.

In his model, Watson saw that each type of base (G, C, A, or T) on one strand of DNA must form a bond with just one type of base on the other strand — with hydrogen atoms providing the bond. According to 'Chargaff's rules', he assumed Guanine would bond only to Cytosine and Adenine bond only to Thymine. He assumed hydrogen atoms would form the bonds between the complementary bases on opposite strands.

Watson and Crick then thought that building a 3-D structure in a helical form of these bases might model a real DNA molecule. They cut out pieces of thick paper (as symbols of chemical organic rings) to arrange them geometrically together as a helix model, holding the cut-outs together with wires.

At first, they had tried a triple-helix model, and it did not work. They were dismayed but persisted. They knew they had the chemistry right. They just needed to get the geometry right. Then, Watson learned through the lab 'grapevine' that Franklin had got a good picture. But Franklin had not shown it to any one yet. She wanted to first analyse the picture. It was her experimental picture, so she had the right to be the first to analyse it.

Watson feared that if he simply asked Franklin to see it she would not show it to him. So Watson did not ask. Franklin was as fierce a scientific competitor as Watson. They were both new post-docs seeking to build a career in science. Both she and Watson were looking for the structure of DNA. Watson went uninvited into her lab and looked at the picture on her desk — without Franklin's permission.

There it was, and Watson saw it! It was clearly the pattern of a helix; but the pattern was of a double helix! Quickly, Watson measured the pattern and rushed to Crick with the information on the angle of the helix. Watson and Crick put their model together in the form of a double helix, two strands of amino acid chains, twisting about each other like intertwined spiral staircases. They used the angle for the helix, as measured

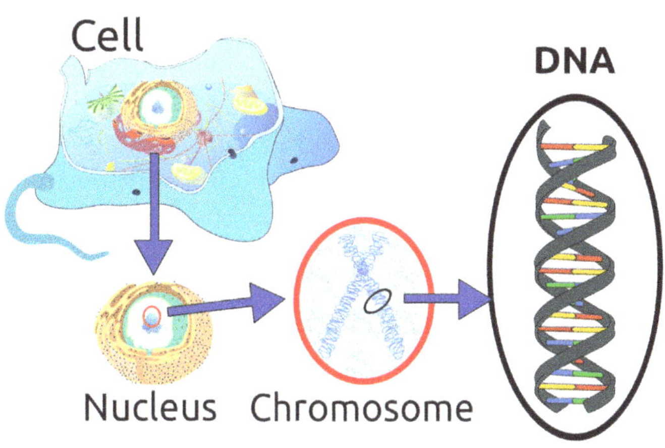

Figure 7.10. The double helix structure of DNA, within a chromosome, in the nucleus of a cell.

(*Source*; WikiiMedia Commons. http://en.wikpedia.org, DNA, 2017).

from Franklin's picture. All the physical calculations and organic chemistry then fit together in the model beautifully! Without a doubt! This was it! This was the major goal of biology: to determine the structure and mechanism of DNA. DNA had a double-helix molecular structure! *This was progress both in scientific epistemology and ontology! The chemical methods of model construction, and the x-ray picture of the molecular geometry together (scientific method) provided the evidence for the nature of DNA (ontology).*

The structure itself was informative about the dynamics of mitosis. DNA was structured as a pair of coiled complementary strands. During replication, the two templates uncoiled and separated from one another, providing two identical patterns for constructing proteins, that is, reproducing life. Through this untwisting and the subsequent chemical reproduction of proteins, life was biologically inherited.

The story of DNA proceeded — but not so nicely in terms of ethics. When Watson and Crick published their paper, they added Wilkins's name onto it, because they had used research from his group, but they did not add Rosalind Franklin's name. That Wilkins allowed his name to appear without Franklin's name and that Watson and Crick did not offer to have her as a co-author — this has gone down in science history as poor professional conduct.

In 1995, Watson and Crick and Wilkins were awarded the Nobel Prize in Biology. Unfortunately, Rosalind Franklin was not honoured. The reason given by the Nobel committee spokesperson was that she had died before the prize was awarded. Yet at the time, the committee could have acknowledged and honoured her major contribution (even without further dividing the prize money). Watson became director of the Cold Spring Harbour Laboratory — *science administration*. In 1988, Watson was appointed head of the Human Genome Project at the National Institutes of Health. Crick contributed to research to try to decode the genetic code and later worked on the idea of consciousness as a biological process.

But some parts of Franklin's story were good. She felt she had not been treated well at Kings College, so she went to France; there, she was well treated and continued to do excellent research in picturing organic structures. Unfortunately, she died young due to cancer. Some have conjectured that it was the exposure to X-rays in her research that triggered the cancer.

This mention of improper acknowledgement of contributions to scientific progress is not meant to denigrate the major contributions of Watson, Crick, and Wilkins. They were great scientists, but they failed to give proper acknowledgement to Franklin. She should have been made a co-author of the paper as she did the actual X-ray research, not Wilkins. Proper acknowledgement of individual contributions to scientific progress is an essential feature of the 'openness' of scientific method. If scientists are not given proper recognition, there is no personal incentive to publish openly and freely. *Science is a kind of managed activity: managed in research projects, managed in research journal publications, managed in professional recognition, and managed in research and teaching. (A managed activity is any organizational process controlled by managers and performed by operations personnel. Science also is performed in organized processes.)*

In summary, all the empirical evidence and theory ready for the construction of a geometrical and chemical model of DNA was not assembled until February 1953, when James Watson and Francis Crick made the first of their geometric models of the structure of the DNA molecule. The nucleotide bases were geometrically flat in a ring structure. Then, one more empirical piece of evidence was needed to get the geometry of DNA correct. This came from Rosalind Franklin's x-ray pictures of DNA. Moreover, the structure itself was informative about its biological process. It indicated the molecular action of DNA in the mitosis of cell reproduction. During cell division, the DNA of the mother cell must divide and replicate into two identical sets of DNA for each daughter cell. The two strands of complementary

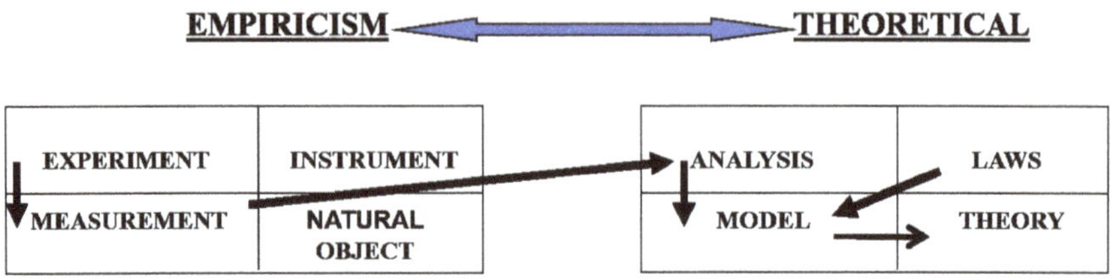

Figure 7.11. Research techniques contributing to the empirical basis for the analysis, laws, and model of DNA in the theory of biological reproduction.

DNA uncoil and each strand duplicates its complementary strand, resulting in two identical copies of the original double-stranded DNA.

In the molecular modelling of DNA, the Experiment to obtain geometrical x-ray Measurements of the DNA crystal were critical to establish the geometry between chemical components (Analysis) to Model DNA as a double helix. Chemical laws were used to construct the model (see Figure 7.11)

(1) *The discovery of the cell and DNA, and*
(2) *The construction of theories of genetics and mitosis, and*
(3) *The demonstration of DNA as the mechanism of hereditary, and*
(4) *The modelling of the chemical molecule of DNA — all these discoveries together formed the theories of molecular biology — the empirically-grounded theory of the transmission of heredity in microbiology.*

Summary

We now return to the table of research techniques to expand on them, Figure 7.12.

Experiment

Experiments are controlled observations of nature. They involve experiencing nature through the human senses aided by scientific instruments. Frederick Griffith's experiment demonstrated

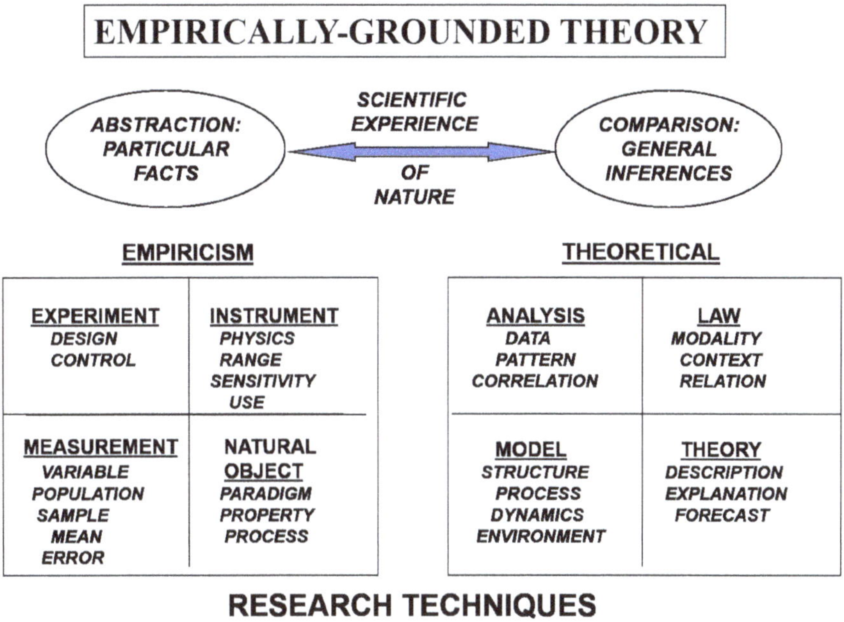

Figure 7.12. Research techniques — methodology of science to ground theory upon experiment.

that DNA was the transmitter of genetic information. Oswald Avery, MacLeod and McCarty reproduced Griffith's experiment, and chemically isolated the transmitting chemical, showing that it was DNA which carried the genetic information. Two important aspects of experimental technique are design and control. *Design and Control are critical features in the conception of the Experiment — determining what scientific question is asked and if it can be answered by data and results obtained from the Experiment.*

Instrument

Instrument is a means of extending the range and sensitivity of human sensing of nature. In the example of the microscope as a scientific instrument one can see that any physical instrument operates by physical interaction with nature. Microscopes use the physics of optical-light waves or X-rays or electron beams to image small objects in nature. The type and range of the physics differs in different instruments. Optical microscopes and X-ray microscopes use different spectra of electromagnetic radiation. A given instrument will detect the physical interaction of nature with a given sensitivity. *In the research technique of Instruments, some important research features of the Instrument are: Physics, Range, Sensitivity and Use.*

Measurement

Measurement is an instrumental way to observe nature expressed in quantitative terms. The quantitative data derived from measurement are both valid and reliable. Measurement is also now a major topic in research methodology, treated from a statistical point of view. Statistics is a mathematical topic concerned with selecting an appropriate sample for the measurement of a population and then estimating the likely value of the measurement for the population from the sampled measurement. Statistical methods enable estimation of the error in the measurement of the population, depending on the size of the sample. *Measurement estimates the quantity of a Variable from statistical techniques about Populations and Sampling and estimates of the Mean value and Error of a Measurement.*

Natural Object

A natural object is an abstraction and generalization of a class of observable things in nature (such as a 'biological cell', a 'molecule', an 'atom' or a 'protein'). Observing and understanding the existence and interactions of a natural object provides knowledge about all the things in nature which exist in this class. The nature of the object must be properly expressed in the right scientific paradigm. For example, physical objects are described as mechanisms, while social objects are described as functions. DNA is a type of molecule essential to reproduction; and RNA is a related type of molecule essential to enable DNA to construct proteins by way of RNA. *A natural object is a natural thing first depicted within a scientific paradigm appropriate to the object's properties and/or processes.*

Analysis

Analysis consists of inferring a mathematical pattern in the quantitative data obtained from experiments. To analyse means to abstract the underlying form of the data, and to generalize the form so that data from additional new observations would fit that form. Analysis of data is the first connection of Empiricism to Theory. Mendel's analysis of the pattern of inherited traits provided the idea of dominant and recessive traits in inheritance. *Theoretical analysis of experiments provides the first step in constructing empirically-grounded theory by determining the qualitative and/or quantitative patterns in experimental Data.*

Laws

Laws are the expressions of the processes discovered in relationships between natural objects. When empirical evidence contributes to the discovery of relationships in nature, such relationships are called phenomenological laws. In the discipline of biology, phenomenological laws were called biological 'principles' instead of 'laws'. The principles of heredity were established by Darwin's epic work on evolution and by Mendel's pioneering work on genetics in the 1800s. In 1910, Thomas Morgan's experiments demonstrated the validity of Mendel's principles, and also showed that mutations could be induced by x-rays, providing a mechanism for Darwin's evolutionary mechanisms. Phenomenological laws can be of different modes, depending on the discipline, such as causal laws in physics and prescriptive laws in social science.

Model

Model is a symbolic simulation of the processes of a phenomenal object. James Watson and Francis Crick constructed a double-helix model of the DNA as two complementary strings of base sequences. The molecular action of DNA in the mitosis of cell reproduction occurs with the two strands of DNA dividing and then each strand duplicating itself. *Models theoretically represent (describe and explain) objects (phenomenological things) observed in experiments carried out on nature.*

Theory

Theory is a symbolic description and explanation of a phenomenal field of objects. The theory of genetic inheritance and the modelling of DNA and its process in mitosis together provided the theory for molecular biology.

Concluding Remarks

The construction of scientific theory involves circularity in inductive and deductive interactions between empiricism and theory over time (sometimes a long time).

1. Scientists pursue research that asks the very basic and universal questions about the existence of things and how they work. (In the case of genetic engineering, the science base was guided by the questions: What is life? How does life reproduce itself?)
2. To answer these questions, scientists require new instrumentation to discover and study things. (In the case of genetic research, the microscope, chemical analysis techniques, cell culture techniques, x-ray diffraction techniques, and the electron microscope were some of the important instruments required to discover and observe the gene and its functions.)
3. These studies are grouped in different disciplines which use different and specialized instrumental and theoretical techniques. Scientists (e.g. biologists, chemists and physicists) undertake these studies. (Even among the biologists, those specialists in gene heredity research differ from specialists in

viral or bacterial research.) Accordingly, science is pursued in disciplinary specialties, each scientist seeing only one aspect of the existing thing. Nature is always broader than any one discipline or disciplinary specialty.

4. Major advances in science occur when sufficient parts of the puzzling object have been discovered and observed and someone imagines how to put it all together properly (as Watson and Crick modelled the DNA molecule). A scientific model is conceptually powerful because it often shows both the structure and the dynamics of a process implied by the structure.
5. Scientific progress takes much time, patience, continuity, and expense. Instruments need to be invented and developed. Phenomena need to be discovered and studied. Phenomenal processes are complex, subtle, multileveled, and microscopic in mechanistic detail. (In the case of gene research, the instruments of the microscope and electron diffraction were critical, along with other instruments and techniques. Phenomena such as the cell structure and processes required discovery. The replication process was complex and subtle, requiring the determination of a helix structure and deciphering of nature's coding.)

CHAPTER 8

Funding Disciplinary Science

Introduction

As we have noted, industry directly uses technology, but not science, to design its new products. It focuses research on applied research and the development of technology — not on basic research such as disciplinary science. In contrast, academic departments in universities focus on basic research within disciplinary boundaries to advance knowledge, instead of targeting research to improve technology. Yet we have seen that (in the case of basic research in the modelling of DNA) over a long period of time, basic research in universities can lead to radical new technology (recombinant DNA). This still leaves the question about how basic research can be used in the short term to advance technology. *Disciplinary basic research is ultimately important to technological progress; yet, it may take many years and much effort to finally assemble enough information about a natural object to know it well.*

Historical Case: U.S. National Science Foundation

Since disciplinary basic research is the strength of university research and the basis for advancing science, we should pause to review how to effectively fund basic research in universities. All nations should have a government agency dedicated to funding science in universities and producing doctoral level graduates. As an example, we will review the operations of the United States National Science Foundation (NSF); a body that is responsible for the funding of about one third of the country's basic research.

Established in 1954 by the U.S. to fund both basic research and applied research (which meant 'science' and 'engineering' research) at American universities, the idea of creating the National Science Foundation (NSF) was conceived by Vannevar Bush, an MIT engineering professor, who ran the U.S. Office of Strategic Defence during the Second World War. He published an important paper on U.S. science and technology policy called "Science: The Endless Frontier." (Bush, 1950) The bill for establishing NSF was signed into law by U.S. President Harry S. Truman.

The NSF headquarters are located at Arlington, Virginia, USA, with about 1700 employees; half of these are PhD scientists and engineers who administer the research divisions and programs of NSF. NSF receives research proposals from professors at American universities and processes these proposals, using peer review to judge quality.

A research proposal is a plan to undertake a finite project of scientific research; it can be submitted to a research funding agency (such as NSF) to obtain resources to perform research. University professors submit research proposals for funding to hire graduate students to undertake research and purchase instruments needed for the research. In 2007, NSF received about 40,000 proposals for research projects, of which about 10,000 were granted research funding. (NSF, 2017)

NSF's proposal selection process is depicted in Figure 8.1. A basic research proposal consists of

- A plan for a research project.
- *A research strategy for the project is described in the proposal.*
- The reason for the research project is to address issues about understanding nature, which is needed for the progress of science. Together
- *research strategy and research issues* are necessary to gain funding for the research — a *research grant.*

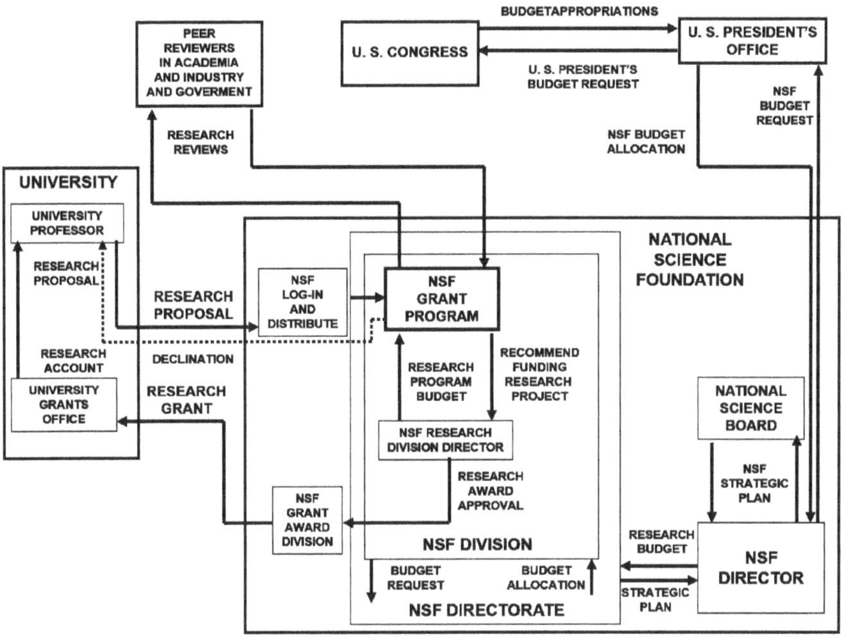

Figure 8.1. The grant operations of the U.S. National Science Foundation.

These research grants support graduate students doing doctorate programs in science and engineering. *This is the science & technology policy to integrate research and graduate education in the U.S.*

The application process begins at a research university, where a professor with an idea for a research project writes a research proposal and submits it to the university for approval. The Chairman of the professor's department and Dean of the department's school must then approve (sign-off) on the proposal, and it goes to the university's grant administration office. When all the proper administrative rules are satisfied, the grant administration office submits the professor's research proposal to NSF. The reason for this formal procedure is that an NSF grant is formally awarded to the university for its administration — the professor is the principal investigator (project manager) responsible for the performance and deliverable of the research funded by the grant.

An electronic research proposal reaches NSF and is logged in and distributed to the appropriate NSF research program. Each research program has a research program office with an expert holding a doctorate in the field of the research/program to handle the review of the proposal. Professional responsibilities of the program officers include:

(1) Reading proposals and identifying the proper kinds of scientific expertise required to review the research quality of the proposal;
(2) Selecting peer reviewers in academic, governmental, or industrial research positions who have published in the areas of expertise indicated in the proposal who are qualified to judge the research quality;
(3) Electronically forwarding a copy of the research proposals to reviewers, with a request to review the proposal, along with proper regard to confidentiality of information in the proposal;
(4) Receiving the reviews and deciding if the proposal should be funded based on advice in reviews, and grading the proposal as excellent, very good, fair or poor.
(5) Allocating a grant budget to proposals rated at excellent and very good, which the program officer has decided to fund (within the program budget);
(6) In some programs, a Program Officer will use an external panel (selected by the program officer to offer funding advice), which will meet under the auspice of the Program Officer and rank in terms of quality the reviewed research proposals, and then the Program Officer, on this advice, allocates funds to proposals (projects or centres) within the program budget constraint;
(7) Forwarding recommended proposals for grants to the Division Director overseeing the program;
(8) Upon approval by the Division Director, the proposals to be funded as research grants are then sent to the NSF Division of Grants in the Directorate of Budget, Finance & Award Management;

(9) The Division of Grants notifies the Professor and University administration (for each approved proposal) that a research grant based on the proposal will be awarded for a certain amount to begin by a certain date;
(10) After formal notification of the award, the University administration establishes a research budget-line upon which the Professor can draw upon, to fund and begin the research project;
(11) As the project progresses, the Professor sends any required progress reports to NSF and the University administration sends requests for reimbursement of research spent on the grant.

In the progress of science, there is direct participation by a program officer (science administrator) as to grant decisions, deciding which projects to fund.

The organization of NSF has been divided into disciplinary research directorates, which has often followed the disciplinary organization of research in universities. In 2009, the NSF research directorates were organized by scientific disciplines:

- Biological discipline (Biological Sciences),
- Computer discipline (Computer & Information Science & Engineering),
- Engineering disciplines (Engineering),
- Earth science discipline (Geosciences),
- Mathematics and physics and chemistry and materials disciplines (Mathematical & Physical Sciences), and
- Social science discipline (Social, Behavioural, & Economic Sciences),

Two other directorates were educational or administrative: (Education & Human Resources), (Budget, Finance & Award Management), and Information & Resource Management).

Each research directorate is divided into divisions (which have often followed the organization of university disciplinary departments), and then into research programs (which approximate disciplinary sub-specialties). Research proposals for research projects are submitted to an appropriate research program. In each research program, NSF defines the focus and scope of research issues within which proposals could be funded, and the criteria used by peer reviewers to assess the quality of submitted proposals. Finally, outstanding proposals are funded from research program budgets.

The NSF budget to fund research comes from an appropriation bill passed by the U.S. Congress and signed by the U.S. President. For the budget process, the NSF directorates submit budget plans to the Director and National Science Board annually. According to the NSF authorization law, the U.S. president appoints both the Director and members of the Board. The Director and Board share equal responsibilities in governing NSF. A preliminary budget request agreed upon by both Board and Director is submitted to the President's Office of Management and Budget (OMB). On examination, the examiner looks at the budget request and discusses it with the Director of OMB to set a budget target for NSF for the next budget year. This target provides an upper bound for the NSF budget. The Director and Board then revise the NSF budget to be within the OMB target and resubmit the budget to OMB.

This process of budget targets and budget submissions occurs between OMB and all Federal agencies. In January, OMB assembles all the budget requests to the President from all the agencies — and submits the whole budget package to the U.S. Congress in February as the President's Budget Request (for the next Federal government fiscal year). From February to October, authorization and appropriation committees in Congress review budget requests. Congress passes 12 appropriation bills for the different federal agencies in order to fund the Federal government for the next fiscal year. This is the formal U.S. process of passing a new annual budget.

When the appropriation bill in which NSF is included has been passed, the Director of NSF assigns budgets to the NSF directorates — which in turn assign budgets to NSF divisions and programs within divisions. The program research budget funds research proposals with research grants.

Peer Review in Research Funding

Research grants support the scientists (professors and graduate students), and provide the equipment to undertake the scientific research planned in the proposal. Research proposals describe the issues of the research to be performed, and the methods and team for performing the research. Research proposals must be sent to an appropriate research program in a government research-funding agency.

> *The reason to begin with understanding the strategy of a government-research-funding agency is this: a research project which is perceived by a program as outside its planned scope seldom gets funded by the program — no matter how good the proposal is.*

The process by which government research agencies fund research proposals from universities is sketched in Figure 8.2. There are two levels in the processing of research proposals: the university and government levels.

At the university level, university professors submit proposals, and when a proposal is selected for funding, the professor receives an award letter. Thereafter, the professor undertakes the research, with some periodic overseeing from the government agency. For example, in research awards made by the National Science Foundation to U.S. university professors, the award is usually for three years, with the funding for the project distributed on a yearly basis. After the first year of the award, a professor sends a request for continuation to NSF with a short report on progress; and the NSF program officer can then approve a second year of funding. After a second year of progress, the professor requests a third year of funding, which can be approved by the NSF program officer. After completion of the project in the third year, the professor submits a final project report to NSF, and submits papers to scientific journals for publication. To obtain more funding, the professor must submit a new proposal.

A project proposal describes the goal of the research in terms of addressing an issue in science or in technology. Proper formulation of a research issue is the first stage of scientific methodology, requiring both the issue formulated to be cognizant of the relevant research literature, and a sound methodological approach. Research is expensive and funding is required to pay the salaries of the researchers, to purchase the research equipment, to maintain the research laboratory facilities and to support research administration.

At the government level, research programs are formulated at a government research-funding agency. These programs specify the areas and issues which research proposals can address

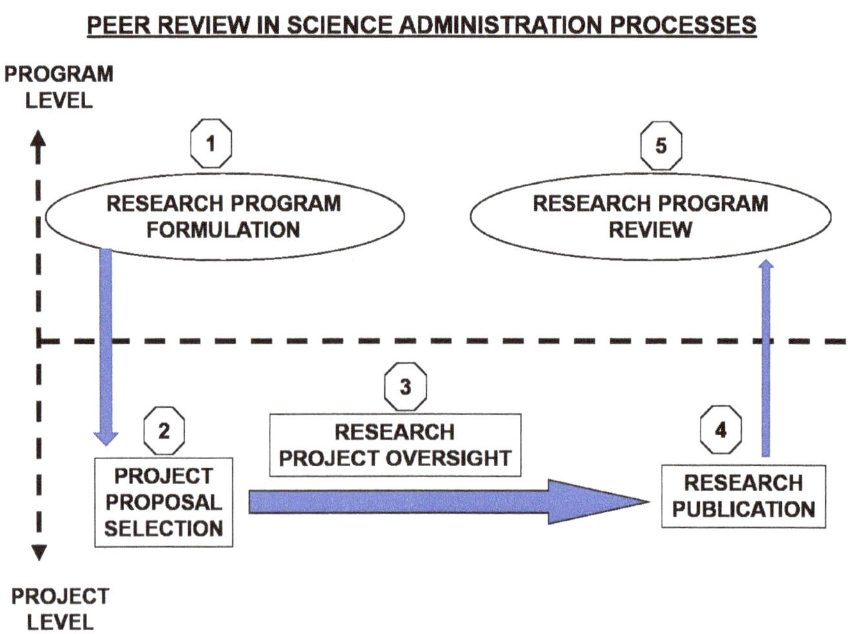

Figure 8.2. Peer review is used in project selection (2), research publication (4) and in research program review (5).

for consideration for funding. A program will formulate what research strategy and methodological approaches are appropriate for the proposals. Periodically, research programs will be reviewed for effectiveness and appropriateness as to strategy.

Within a government research-funding agency, research proposals are reviewed for quality. The funding program appoints an external group of relevant scientific and technical peer-reviewers to read the submitted research proposals, and asks them to grade each proposal in terms of quality as excellent, very good, fair, or poor. The funding agency then selects only among the excellent or very-good rated proposals for funding within their research budget constraints. Proper peer-review is an organizational procedure vital to ensuring the quality of the research methodology of a proposal.

The results of the completed projects should be published in the scientific, technical and professional journals (with possible delays for the filing of patents on intellectual property created in the project) as outputs or deliverables. Scientific, engineering and professional societies publish journals that archive progress in science, engineering and technology. In their journals, these societies also use peer review to judge whether to publish a paper as a contribution to knowledge. Much modern research is performed in discrete projects, which are managed both organizationally and methodologically to maintain its quality.

In 2017 for example, research in Stanford University had an annual cost of about $200 million dollars, with funding from U.S. government agencies (such as NSF and DARPA), philanthropic organizations (such as the Rockefeller Foundation) and industrial firms (such as IBM or Google).

Format of Research Proposals

From the view of the 'philosophy of science', basic research projects provide the format for scientific 'inquiry' — asking and answering a question about nature, using the scientific method. A basic research proposal articulates this approach — asking a finite and specific question about nature. The methodology for answering the question about nature will use at least one of the research techniques discussed in the previous chapter.

In a proposal, the question asked about nature is the *research issue* of the proposal. The way the question will be answered is the *scientific method* of the proposal. The *research impact* of the proposal is concerned with — to whom the answer to the question will be of value.

A scientific research proposal for a research grant should clearly pose:

- *A scientific or technical vision of a research issue, (Originality)*
- *How the research can be performed based on certain research methodology (Ability to test ideas), and*
- *Potential use of the research results and research impact (Contribution to knowledge).*

The format of a basic research proposal (written to gain funding for a scientific project) needs to address four criteria: research vision, research methodology, research team, and research impact.

1. *Research Issue — Vision. (Originality)*
   a. What is the frontier-of-science and/or state-of-art-in-technology in the specific research field?
   b. What are the issues or challenges or barriers to scientific/technical progress in the field?
   c. Which of these issues/challenges/barriers (research issues) will the research project address?
2. *Research Methodology. (Ability to test ideas)*
   a. What research techniques will be used to address these issues?
3. *Research Team. (Ability to test ideas)*
   a. What kinds of skill sets will be required to carry out the methodology?

   b. Which members of the research team will provide these competencies?

4. *Research Equipment. (Ability to test ideas)*

   a. What research instruments will be needed to perform the research?
   b. How will these be accessed or purchased?

5. *Research Impact. (Contribution to knowledge).*

   a. How would the successful results of the proposed project contribute to the advancements of knowledge and/or used to solve important problems?
   b. How will such progress be communicated or implemented and for whom?

6. *Research Budget*

   a. What funding amount is requested for the project and for how long?

University Procedures in Research-Proposal Writing and Submission

When writing and submitting research projects, university researchers consult both scientific and technical journals as well as their university science departments to generate a set of potential research issues that guide future research, as sketched in Figure 8.3.

The scientific techniques of *observation & experimentation & instrumentation,* and *analytical techniques*, as well as *theory & model construction* are all the key foci of research projects. They are used to *discover and understand nature*.

These ideas are used procedurally to formulate *research issues* for a *research proposal*.

But the formulation of the *research issues* also requires information input from *scientific journals* to identify if the issue is novel and significant to future scientific progress. In addition, information input also must come from *funding- program*

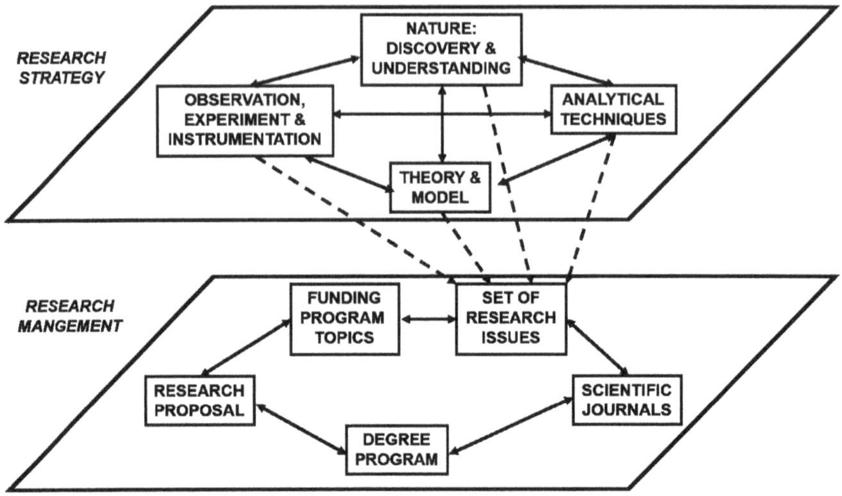

Figure 8.3. Procedures in research management for developing research strategy.

topics to identify a funding program in whose domain the research issue recognizably falls. And information input must also come from a university graduate *degree program* so that the research issue could provide a thesis topic for a doctoral student who would undertake the research tasks as part of the proposed research proposal.

An essential feature for both a research proposal and a research publication is an adequate literature review in scientific journals of all published references up to the frontiers of science. Scientific journals publish and archive the latest research results from recently completed research projects. Consequently, scientists read scientific journals to identify research gaps, so that they can claim originality.

In the disciplines of science, the procedures of journal publication and on-line Internet publication of journals are central to science administration. The journals define their focus on a specific area of science, a small piece of the totality of science. More than one journal may cover the same area or overlapping areas. Journals are usually published by scientific societies, which are voluntary organizations. If there is more than one society covering an area of a scientific field, then there may be more than one journal in the area. For example, there are often national scientific societies and international scientific societies whose interests overlay scientific areas.

Scientific journals use scientists to review the quality of a proposed article from the perspective of methodological soundness and novelty. Scientific journals also serve an archival purpose, permanently recording the results of research projects that have been undertaken and published in recent years. Since research normally takes about three years to perform, at least one year to get funded, and about a year to get published, research ideas in a current issue of a scientific journal are usually five years old. Because of this time delay between scientific conception, achievement and communication, scientific researchers also maintain a more current mode of communication amongst themselves in the form of annual scientific conferences and informal discussion networks.

Accordingly, the research proposal to define the research issue of the proposed project should identify all the relevant recent articles in scientific journals that together describe the frontier of science (or state-of-art of technology) in the area of the research, and the challenges of advancing knowledge, which the proposed research will address. The most significant and relevant scientific journal articles and recent conference presentations need to be identified in research proposals so that they stand out as original.

Once the originality of the research proposal has been established and justified by the review of relevant scientific literature, the next section of the research proposal should describe the research methodology, or how the research will be performed, collectively known as "ability to test ideas". All scientific methodology should contain the procedures for observation and experiments carried out on nature, as well as the procedures for collecting and analysing the results of experiments so as to construct a theory or model. New experiments should then be performed, and data from those experiments used to confirm that the theory or model is accurate.

The scientific method defines the approach of any research strategy. In a research project, the reality of nature and problems involving nature (technical and social) are queried as a research issue in the project. The method of query can be through:

(1) Providing new observations/measurements/experimentation of nature and/or new scientific instrumentation;
(2) Providing new analytical techniques or using existing analytical techniques on new domain;
(3) Providing new theory and/or new models of objects or using existing theory/models in new areas.

Research proposals receive high review ratings and are funded when

(1) *The research issue in the project is seen as being on the frontier and important,*
(2) *The proposed research is novel,*
(3) *The methodology is seen as likely to discover new science, and*
(4) *The research investigator is seen as qualified to conduct the proposed research.*

Particularly, the set of scientific issues must be recognized as within the scope of a research-funding program in a funding agency for the proposal to be seen as valid and timely. Again, the use of scientific journals in defining the frontiers of scientific research is fundamental both to the scientist proposing the research, and to the program officer in the funding agency to which the proposal is submitted. And as we saw in the NSF funding chart, the funding agency program officer will send out the research proposal to scientific peer reviewers to judge the appropriateness of the research issue and the soundness of the scientific methodology to be used in the proposal.

Next, the proposal must describe the research team, their appropriate specialties and their qualifications to undertake the research. Peer reviewers used by the funding agency will also judge the competency of the proposed research team.

Any appropriate research equipment and instrumentation needed to perform observations, measurements, or experimentation in the project must also be described in the research proposal. The software or computer capability needed to develop or use theory or modelling in the research project must be included. All these will also be judged by the funding agency's peer reviewers as readiness and ability to test ideas.

In some scientific projects, the research issue is defined, not only by the state of knowledge (being on the frontiers of science or using state-of-art of technology), but also by some other important problem (encountered by industrial technology or government technology, or a societal problem). In such cases, the research proposal needs to not only address the research issue, but also the impact from the research, and how the results from the research can be implemented.

Finally, the research proposal needs to specify a project duration and research budget for undertaking the research. Research proposals for scientific research at universities normally specify a duration of three years. This period corresponds with the time normally required to do a doctorate dissertation. In universities, graduate students working under a professor as a thesis supervisor perform scientific research in a 3 years PhD program. The scientific workers are the doctoral graduate students, who need financial support for living and tuition expenses. The research grant will provide the funding needed to support them to participate in the project. The number of graduate students working on a research project should be determined by the number of distinct tasks defined in the research proposal, each of which could lead to a doctoral thesis. The cost of the research assistant's salary plus tuition, multiplied by the number of students, equals to the labour cost for the research. In addition to the 'research labour' cost on the project, there is an overhead charge by the university for providing space and other support functions to the project.

> *A research proposal summarizes a possible contribution to science as a discrete future research project.*

Before submitting any research proposal to a funding program, one needs to first understand what the real research focus and boundaries are in the current operation of a program. This can be achieved by finding out the current program officer's perception of the research scope for the program. As the program officers in charge of administering a research program are replaced by others, the program's priorities in research support also change. Some ways to understand a program's research priorities include:

(1) looking up a program's list of recent research grants,
(2) visiting the current program officer,
(3) learning about a program officer's resume and publications.

> *Be creative in formulating a good research proposal but also know the program officer to anticipate the relevance of your research proposal to program priorities.*

Summary

Disciplinary-organized science progresses by individual basic research projects, which add to science (sort of like building a wall, brick by brick) until a synthesizing idea that connects them together explains the puzzle of nature (sort of like placing a keystone in the arch in the stone wall). Individual basic research projects provide the basis for a dissertation by a PhD candidate (who performs the research in the project). Academic basic research projects and PhD education are tightly connected. This is why disciplinary research dominates in university research — to combine scientific progress with graduate education.

In 2017, NSF was divided into seven research funding directorates: Biological Sciences, Computer and Information Science and Engineering, Engineering, Geosciences, Mathematical and Physical Sciences, Social, Behavioural and Economic Sciences, and Education and Human Resources. Each directorate was headed by an assistant director reporting to the director of NSF and NSF's National Science Board, and each is further subdivided into divisions like materials research, ocean sciences and behavioural and cognitive sciences. NSF was created as a government agency to fund research in the science and engineering disciplines, as organized in American universities. It was not founded to support basic research targeted to industrial technological needs.

Disciplinary basic research provides the 'core competence' of research universities. Accordingly, university research capabilities are often narrow but deep. Still, when enough disciplinary research projects get done to solve a puzzle of nature, all that disciplinary research pays off — as 'new technology' emerges, paving the way to welcome 'new industries'.

But in the short run and in the pressing need for technological progress, can disciplinary research be drawn upon for more immediate progress in science and in technology? Yes, and the way to do this is through a complementary type of research centre in a university — a multidisciplinary and industrial cooperative centre. And we review these complementary university centres in the next chapter.

CHAPTER 9

Targeted Basic Research Cooperation: University & Industry

The issue in making scientific progress immediately relevant to technology development is this. How can industry and universities cooperate to perform basic research targeted to industrial needs? For this, universities need a multidisciplinary research center to perform targeted basic research. We will illustrate such multi-disciplinary & cooperative research centers with two programs run by the U.S. National Science Foundation (NSF). These are the Engineering Research Centers Program (ERC) and the Industry/University Cooperative Research Program (IUCRC). Their role in university-industry research cooperation was to facilitate the basic research needed to innovate the next generation of technology.

Next Generation Technologies

We recall that progress in technology can occur incrementally or radically. If *performance* refers to how well technology performs a function, and *function* refers to how technology is used, then incremental innovations create small improvements in the performance of a technology, but do not necessarily change the function of the technology. On the other hand, radical innovations create both large increases in a technology's performance and changes in function. Radical innovations occur periodically, while incremental innovations occur more frequently.

Industries often know how to incrementally develop new technology for its needs, but do not always know how to create radical changes. Industry needs science:

(1) When new basic technologies must be created from new science, and
(2) When technological progress in an existing technology cannot be made without a deeper understanding of the science underpinning the technology. It is when industry needs radical innovation that major problems in the R&D infrastructure arise. Then, industry-university-government research partnerships are especially important.

As every industry faces competitive conditions in the long term, a firm needs to be capable of responding to both incremental and radical changes in technology. Having studied some companies using technology leadership to create a long-term competitive advantage, Joseph Morone (1993) wrote: "The similarities among the cases of success begin with the way general management defines the domain in which their businesses compete. In every case, a strategic focus has been consistently articulated and applied for decades." A long-term business strategy can be constructed on a core technology competency, which leads to product quality and innovation.

As an example, Corning was a traditional glassmaker which moved into making fibreoptic cables in the second half of the twentieth century. Morone (1993) cited Corning's technology strategy in the second half of the twentieth century: "Creating entirely new businesses through radical innovation is a hallmark of Corning's history. The optical fibre business is the most dramatic recent example . . . radical innovations were followed by generational innovations, and within each generation by incremental improvements. In optical fibres, Corning made its own manufacturing capability obsolete by developing and introducing seven distinct generations of processing technology over the course of 15 years."

The importance to industry of a cross-disciplinary university research centre is that science can be focused on targeted basic research to create 'next generation' technology (NGT) for industry. This is a radical kind of innovation for replacing an existing technology, in contrast to the incremental steps taken to improve an existing technology.

Next-generation-technology (NGT) research planning is an excellent way for industries to utilize university-based science. It improves the management capability to exploit technology discontinuities. For technology transfer between industry and university to be effective, it is necessary for both the strategic business units' research labs and corporate research to cooperate with university researchers in formulating and reaching a consensus on an NGT vision and research plan. Industry-university research cooperation can facilitate good cooperation between corporate and divisional research labs to support management with its long-term research needs.

Historical Case: MIT Biotechnology Process Engineering Centre

An example of an ERC multidisciplinary university research centre for biotechnology research is the MIT Biotechnology Process Engineering Centre (BPEC), established under an NSF/ERC grant in 1985. At that time, one of the important research needs of the new biotechnology industry was to make the shift from using *E. Coli* cells to using mammalian cells in therapeutic protein production.

We recall that Boyer and Cohen experimented on *E. Coli* cells to discover the recombinant DNA technique. Recombinant DNA is the technology of DNA alteration which revolutionized the biological science and technology industry. However, individual prokaryotic *E. Coli* cells have no means of secreting out the therapeutic proteins which their modified DNA produce. To obtain produced therapeutic proteins from the prokaryotic *E. Coli*, biotech firms had to lyse the *E. Coli* such that the therapeutic proteins could be isolated. It would be more efficient to produce the modified-DNA proteins in multicellular eukaryotic cells, which would secrete the manufactured protein. Mammalian cells are multicellular eukaryotic cells that can do this because they live together in interactive colonies, and need to send proteins to each other within the colony. Hence, biotech firms will not need to lyse the production units (cells), and only need to isolate the secreted proteins from their biotech processing systems (the production cells growing in solution). *E. Coli* cells as production units for the biotechnology industry were a first-generation technology, and their use was followed by the usage of mammalian-cells as second-generation technology.

The BPEC centre was focused on researching how mammalian cells could function as a biotechnological processing system. This science would enable the biotechnology industry to move from using *E. Coli* cells to mammalian cells in producing therapeutic proteins — an immense improvement for the biotechnology industry. BPEC was founded and directed by Professor Daniel Wang. In this centre, university and industrial researchers planned basic research for the technologies of producing biotechnology products grown in mammalian cell cultures. It has successfully operated for over two decades. (BPEC, 2005) We look at its research plan as originally formulated in 1991 (Betz, 1993).

The technology system for biotechnology production divides into a *bioreactor subsystem* (which grows the cell cultures), and a *recovery subsystem* (which recovers desired proteins produced by the cells).

Within the bioreactor, the technological performance variables for protein production are:

- Number of cells
- Performance of each cell
- Quantity of the protein produced by the cells
- Quality of the protein produced by the cells

The 'products' of biotechnology cell production are therapeutic proteins that have commercial use in medicine. The quantity of protein is a measure of productivity. The quality of proteins is affected by the construction and folding of proteins. Proteins are long chains of organic molecules that fold back into particular configurations. The usefulness of the protein depends not only on having the right molecules link together, but also on its ability to fold up into the right structure. One of the technical bottlenecks in protein production was to get the proper folding of DNA after its secretion from a cell.

Biotech engineers needed to design bioreactors for protein production from mammalian cells that could:

- Grow a high density of cells
- Formulate the proper medium for their growth
- Provide proper surfaces for mammalian cells to attach to
- Control the operation of the reactor
- Provide for the recovery of the proteins that are grown in the cells

In the design of the recovery processes of the protein product from the bioreactors, biotech engineers needed to control the physics and chemistry of protein structures, including:

- Protein-protein interactions that avoided the aggregation of proteins or the mutations of proteins
- Protein-surface interactions that affect the aggregation of proteins or the adsorption of proteins to separation materials
- Protein stability that affected the durability of the protein through the separation processes.

Planning Targeted Basic Research

For either an ERC or IUCRC centre, the proper leadership of a centre director is critical to its performance. The two basic qualifications for a professor to become a university research centre director are:

(1) to have a scientifically exciting multidisciplinary research vision, and
(2) to successfully win a research grant for a centre from industries and/or government.

The strategic responsibility of a centre director is to have a research vision, which can be realized through the planning, conducting and assembly of discrete research projects in the centre — integrated research, which together will coherently yield integrated scientific and technological progress. A strategic, multidisciplinary research plan for a university should be expressed as sets of projects grouped on three planes of research goals:

(1) A science base,
(2) A set of new technology sub-systems, and
(3) Experimental test-beds for prototypes of a new technology system.

A winning centre proposal requires that the group of researchers around the proposed director envision particular exploratory projects which could open up a new technological area. The federal government, in collaboration with industry and university researchers, can organize a national centre to identify new opportunities for research. *If the government and industry subsequently fund such a university cross-disciplinary centre, projects can be monitored to identify both the direction and pace of technical advances.*

Research planning for the long term should try to envision and research discontinuities as they might be the next generation of the technology. The first step in research planning for an NGT system requires a vision.

A vision for an NGT requires not only the envisioning of a major improvement in performance, but also research imagination:

1. A description of the functionality, performance, and features that are highly desirable, but that cannot be obtained using current technology
2. Imagination of how current applications could be better accomplished with the new technology, and how possible new applications, which cannot be accomplished with current technology, could be accomplished with the new technology
3. Some new methodological research advances that might be used in a new generation of the technology

A vision of an NGT is a description of the functionality, performance, and features desirable that cannot be obtained in the current generation of technology. It includes the imagination of how current applications could be better accomplished with the new technology, and how new applications which cannot be accomplished with the current technology might be feasible with the new technology. A research strategy leading to a technical feasibility demonstration of an NGT consists of a set of interdisciplinary research trusts containing projects scheduled for integration into an experimental test-bed embodying the functionality of an NGT system.

Because the form of an NGT system is complex, one should try an example of a system on an application in order to see which advanced research ideas will be useful, which ones will not, and what cannot be done yet. A research test-bed for an NGT system allows the demonstration of advanced ideas. These reduce risk by making clear what is the most useful engineering pathway to the future generations of a technology system.

Historical Case Continued: MIT Biotechnology Process Engineering Centre

At the MIT BPEC ERC sponsored by NSF, the Centre director sketched out the research vision for replacing *E. Coli* cells with mammalian cells for the biotechnological production of therapeutic proteins. This was expressed as a system analysis of the bioreactor and the recovery system of biotech processing. Next, the centre planned the needed research to implement this vision in terms of scientific issues about mammalian cell cultures which must be discovered.

For the bioreactor portion of the system, the centre listed the following scientific phenomena as requiring better understanding:

1. Extracellular biological events
   - Nutrition and growth factors of the cells
   - Differentiation factors of the cells
   - Redox (reduction-oxidation) oxygen conditions in the cellular processes
   - Secreted products of the cells
2. Extracellular physical events
   - Transport phenomena of materials in the medium
   - Hydrodynamics of fluid-gas interactions of the medium
   - Cell-surface interactions
   - Cell-cell interactions
3. Intracellular events and states
   - Genetic expression of the proteins in the cells
   - Folding of the proteins and secretion from the cells
   - Glycosylation of the proteins
   - Cellular regulation of secretion
   - Metabolic flows in the cells and their energetics

This is a list of the scientific biological and physical phenomena underlying the cellular activities in the technology of the bioreactor, i.e. the science underlying a technology.

For the recovery process of the system, the centre identified the scientific phenomena which required better understanding:

1. Protein-protein interactions
   - Aggregation of proteins
   - Mutations in protein structure
2. Protein-surface interactions
   - Aggregation of proteins through denaturation
   - Adsorption of the proteins to surfaces
3. Protein stability
   - Surface interaction
   - Chemical reaction
   - Aggregation in the solvent
   - Stabilization

Accordingly, in 1992, the Centre organized its research into two areas:

- Engineering and scientific principles in therapeutic protein production
- Process engineering and science in therapeutic protein purification

In therapeutic protein production, the research projects were:

- Protein trafficking and posttranslational modifications: glycosylation and folding
- Redox potential
- Pathway analysis
- Intercellular energetics
- Regulation of secretion
- Hydrodynamics: gas sparging
- High-density bioreactor designs

- Substrata morphology for cell attachment
- Expression: transcription factors

The research projects in the second research area of protein separation were:

- Protein adsorption: chromatography and membrane
- Protein aggregation
- In vivo protein folding
- Protein stability in processing, storage, and delivery

Daniel Wang was founder and director of the centre from 1985 to 1995. Daniel Wang was a professor in the Chemical Engineering Department at MIT. He obtained his bachelor-of-science in biology from MIT in 1959 and a master's degree in chemical engineering in 1961. In 1963, he obtained his Ph.D. in chemical engineering from the University of Pennsylvania. His research interests were in bioreactor engineering and the production of recombinant proteins. He was a member of the U.S. National Academy of Engineering since 1986. (The DIC Wang Group, n.d.)

As of 2017, the Biotechnology Process Engineering Centre is still in operation. Having succeeded in accomplishing a scientific understanding of the biotechnology of mammalian bioproduction systems, the Centre's research program evolved to focus on gene therapy: "Our strategic plan continues to identify selective delivery of gene therapeutics as the crucial bottleneck problem of the envisioned gene therapy. This view remains strongly consistent with consensus in the field to focus on two complementary classes of ultimate engineered gene delivery systems as driving our integrated research, education, and industry interaction efforts:

(1) An ex vivo approach employing genetically engineered stem cells (Project Area A)
(2) An in vivo approach employing targeted viral or synthetic vectors (Project Area B)

Our vision is to generate, as novel engineered systems, prototype selective gene delivery vehicles in each of the gene therapeutics categories still requiring major research advances in order to achieve appropriately effective gene delivery vehicles.' (BPEC, 2005)

Research Strategy for a Next-Generation Technology Vision

Once an NGT vision is created, the next step is to formulate a long-term research strategy:

1. The current state of the technological system should be described, delineating present limitations on functionality as performance limits and existing features and current costs, along with present applications of the technology.
2. A desirable significant change in performance and features is then planned, which would constitute a dramatic leap forward, to alter existing markets and create new markets. The NGT goals are then expressed in the form of planned performance, features, and cost.
3. The research issues which could lead us from the present state of technology to the realization of the NGT vision are identified. These issues are then grouped into a set of interdisciplinary research thrust areas which together constitute a program of basic science and engineering research for the next generation of the technological system.
4. Initial research projects are proposed in these thrust areas to address a range of research issues to be coordinated in time into the prototype subsystems of the NGT system. A milestone chart is formulated, which plans the integration of these prototype subsystems into an experimental testbed for the NGT system.

These steps lay out a targeted basic research program aimed at demonstrating an experimental prototype of the NGT system. The steps are effective because all technologies are 'systems', and the morphology and logic of technology systems can be described.

Industry can identify technology bottlenecks in current technology systems and organize research programs in industrial laboratories to address these bottlenecks. Individual research projects in such a centre can perform basic research focused on improving a technology system through inventing improvements in any aspect of the system:

1. Improved system boundary
2. Improved components
3. Improved connections
4. Improved materials
5. Improved power and energy
6. Improved system control

The natural phenomena underpinning a system can be focused on any of the system aspects:

1. Phenomena involved in the system boundary
2. Phenomena underlying components
3. Phenomena underlying connections
4. Phenomena underlying materials
5. Phenomena underlying power and energy
6. Phenomena underlying system control

Since science focuses on phenomena, targeted basic research is scientific research motivated by any of the above.

In summary, targeted basic-research planning should:

1. Characterize a technology as a system
2. Identify technical bottlenecks in the way of improving the system
3. Imagine if and how new research instrumentation, instrumental techniques, algorithmic techniques, or theoretical modelling could be focused on the bottlenecks.

4. Imagine how to improve the understanding and manipulation of the phenomena of the bottlenecks, or to develop alternative phenomena to substitute new technical manipulations at the bottlenecks.

Successful targeted basic research can then result in a next-generation of a technology system to dramatically jump technological progress

NSF ERC Program

The MIT BPEC centre was one of the first of the NSF ERC centres, begun in 1985. This centre and other early ERCs demonstrated that such multidisciplinary centres could make important research contributions to develop radical innovations that resulted in next-generation technologies. Such ERC-type centres span both science and engineering disciplines relevant to next-generation technologies. They are guided by a research vision for simultaneous progress in both science and technology, towards the next-generation of a technology. About its engineering research centres, NSF (n.d.[a]) had posted:

> "The goal of the ERC Program is to integrate engineering research and education with technological innovation to transform national prosperity, health, and security. ERCs create an innovative, inclusive culture in engineering to cultivate new ideas and pursue engineering discovery that achieves a significant science, technology, and societal outcome within the 10-year timeframe of NSF support. Those who submit proposals in response to this solicitation will need to address the following questions: What is the compelling new idea and how does it relate to national needs? Why is a centre necessary to tackle the idea? How will the ERC's infrastructure integrate and implement research, workforce development and innovation ecosystem development efforts to achieve its vision?"

In 1994, NSF published an interim evaluation of the ERC program: "In 1994, the NSF's Engineering Education and

Centers (EEC) Division initiated study of key aspects of the ERC Program: ERC-industry interaction and the effectiveness of former ERC graduate students in all sectors of employment. The primary intent of this *two-part study* was to examine the extent to which the Program was making progress toward its goals. The purposes of the ERC-Industry Interaction study were to (1) examine the patterns of interaction that have emerged between ERCs and industry; (2) determine which types of interaction were most useful to industry and brought firms the greatest benefits; and (3) assess the value of these to the companies." (Parker, 1997)

In the first ten years of the program, the percentage of industrial sector participation varied:

21.6%	Electrical, electronic, telecommunications
15.8%	Chemicals and allied products, including pharmaceuticals
9.1%	Surgical/medical instruments and supplies
6.7%	Petroleum refining and extraction
4.4%	Aircraft and missiles
3.5%	Machinery
3.5%	Building construction
3.2%	Motor vehicles and motor vehicle equipment
2.9%	Primary metals
2.6%	Professional and scientific instruments
9.4%	Other non-manufacturing (e.g., engineering design/ service and computer software companies)

With regards to industrial participation, the study emphasized: "At the core of the ERC Program is the concept of firms having membership in a center, which involves them in strategic planning, joint research projects, mentoring of students, access to the center's facilities and equipment, access to center researchers and students... Center membership usually involves payment of a fixed annual fee that is pooled with cash from other members and sponsors for support of the center's research and research-related activities... Centers set their own membership rates and often have associate memberships for small firms that cannot afford the cost of full membership... Member companies may also sponsor specific research projects and some form consortia around larger projects and testbeds" (Parker, 1997). The ERC program in NSF has continued since 1985 to the present time of 2017.

The NSF Engineering Research Centres Program was in the Engineering Education and Centres Division, within the Engineering Directorate. A list of recent ERC awards can be found at the NSF website.

NSF IUCRC Program

Earlier in 1963, even before the NSF/ERC Program, NSF had established a university-industry cooperative centre program to encourage industrial participants to fund research in universities — the Industry University Cooperative Research Centres Program (IUCRC). In this program, NSF provided a grant to a university to establish a cooperative research centre, to be used only for a centre's administrative expenses, and the industrial members of the IUCRC centre fully funded the research. The industrial members formed a board to manage the centre, and, with the director, set budgets and selected research projects for the centre.

The difference between the NSF Engineering Research Centres Program and the NSF IUCRC Program is that NSF provided major government funding in ERC centres, and only nominal government funding in IUCRC centres. Thus the scale of the research effort in ERCs was usually many times larger than that of IUCRCs. Also, IUCRCs were governed by an industrial membership board; ERCs were not directly governed, only advised by industry.

NSF (n.d.[b]) published: "The National Science Foundation invests in these IUCRC partnerships to promote research programs of mutual interest, contribute to the nation's research infrastructure base, enhance the intellectual capacity of the engineering or science workforce through the integration of research and education, and facilitate technology transfer... The IUCRC program seeks to achieve these goals by:

- Contributing to the nation's research enterprise by developing long-term partnerships among industry, academe, and government;
- Leveraging NSF funds with industry to support graduate students performing industrially relevant pre-competitive research;
- Expanding the innovation capacity of our nation's competitive workforce through partnerships between industries and universities; and
- Encouraging the nation's research enterprise to remain competitive through active engagement with academic and industrial leaders throughout the world."

An example of a very successful NSF IURC was the Intelligent Manufacturing System Centre. Jay Lee at the University of Milwaukee founded this IURC, and he innovated a new scientific approach to the technical challenge of diagnosing machine wear and maintenance in manufacturing factories. The Centre (2018) published: "IMS is internationally recognized as the leader in predictive analytics and industrial big data modelling for life cycle performance of industrial systems... It is a leader in the discovery of new methods to assess machine degradation and predict the health of industrial systems including e-manufacturing, e-maintenance, cyber machine systems, cloud-based machine monitoring and manufacturing, intelligent cyber machine systems, etc. These discoveries translate into a useful methodology and tools for the development and deployment of prognostics and health management of industrial systems. The Center envisions the future of maintenance as a system that enables equipment to achieve and sustain near-zero breakdown performance with self-maintenance capabilities (self-aware, self-predict, self-compare, and self-configure), and ultimately to realize the autonomous transformation of raw data to useful information for improved reliability, productivity, and asset utilization."

The Centre had been directed by Professor Jay Lee at the University of Cincinnati: "Dr. Jay Lee is Ohio Eminent Scholar, L.W. Scott Alter Chair Professor, and Distinguished Univ. Professor at the Univ. of Cincinnati and is founding director of the National Science Foundation (NSF) Industry/University Cooperative Research Centre (I/UCRC) on Intelligent Maintenance Systems (IMS, 2018)

The Centre began in 2001 and was still operating in 2017, then having over 90 global companies as members. Professor Lee also served as an advisor to several companies, including McKinsey, Alstom Transport, France and GE European Digital Foundry. He also served on the boards of several start-up companies, including Predictronics and Falkorny. (IMS, 2018)

The IUCRC research by Professor Lee and his students created the scientific/engineering basis for a new approach to machine maintenance. Sensors detect signals from the vibrations and sounds of machine operations and research by the IUCRC students translated these signals into functional information. When should tools be changed, or when should bearings be lubricated? Anticipating machinery breakdowns and preventing them would save costs in manufacturing factories. Many of the IUCRC companies began new businesses in predictive machinery maintenance on the basis of the new next-generation technology.

Research Cooperation in Next Generation Technology

Research cooperation is essential to the speed at which an industry in a nation achieves the competitive next generation of a technology, or in other words, to rapid innovation. Cooperation is necessary because of the different conditions for research in industries and universities. Industrial researchers are very sophisticated about current technology and are particularly knowledgeable about its problems. Industrial researchers can easily identify the roadblocks to technical progress. However, because of the applied and developmental demands on industrial research, they usually have limited time and resources to explore ways to surpass current technical limitations. In contrast, academic researchers have the time and research students to explore fundamentally new approaches and alternatives that may easily surpass existing technologies — if funding could be provided to study the problem.

Together, industry and university researchers can envision how to effectively bound a technological system in order to envision a next generation of technology. This boundary is an important judgment, combining judgments

(1) On technical progress and research directions which together might produce a major advance and
(2) In the domain of industrial organization that such an advance might produce a significant competitive advantage.

To encourage industry and university research cooperation on NGTs, a bridging institution is necessary — because industries and universities live in almost completely different universes. The industrial universe emphasises technology, usually with a short-term focus, profitability, and markets. In contrast, the university universe focuses on science, taking a long-term view to acquire knowledge that helps to explain the problems.

A cross-disciplinary & cooperative university research centre should connect these two world views, creating a balance between

(1) Technologically-pulled research and scientifically-pushed research,
(2) Short-term and long-term research focus,
(3) Proprietary and non-proprietary research information, and

(4) Vocationally-relevant education and professionally-skilled education.

These are the issues inherent in industry and university cooperation. Properly handled, these provide creative tension:

(1) Linking technology and science in real-time operation, and
(2) Creating progress in knowledge and developing the technological competitiveness of nations.

The government role in NGT is to assist in funding cross-disciplinary cooperative research centres at universities, which cooperate with industries.

University Requirements for Industrial Cooperation on NGT

There is a complementary set of requirements that the university must satisfy for an industry-university research centre to operate:

1. The university needs to provide the environment and contiguous space for the centre and arrange for state-of-the-art instrumentation so that the centre will be on the cutting edge of science and technology.
2. The university needs to provide the personnel, and have recruitment, tenure, and promotional policies that reward and properly balance faculty contributions to both scientific achievement and technological progress.
3. The centre needs to integrate into a multidisciplinary entity that is complementary to the departmental educational structure of the university.
4. The centre needs to train staff for the industry by actively coordinating the educational requirements of programs and students with research requirements of industry.

5. The centre needs to formulate research projects that result in the publication of open literature.
6. The centre needs to inform faculty and students of how to cooperate with industry while still fulfilling the principal missions of the university in education and research.
7. The centre needs to work with both large and small firms, with larger firms providing the principal industrial support.
8. The centre needs to be capable of formulating strategic research planning.
9. The centre needs to be capable of undertaking scientific research at the systems level of a technology.
10. The centre needs to have a dual-structure organization; providing capability for performing scientific research on the one hand, and delivering the technology developed to the industry on the other.

Requirements for Industrial Use of University Science for NGT

For an industry to effectively support and use university research centres, several requirements must be met:

1. A diversified firm must have a corporate research lab, as well as divisional labs.
2. The corporate research lab must be tightly linked to the divisional labs.
3. Such centres should link with both the corporate research lab and the divisional labs of a company.
4. The company should work hand-in-hand with multidisciplinary strategically-focused university research centres.
5. Such centres must be large enough to perform a critical mass of research useful to industry.
6. Company personnel should participate in both the governance and research of the university centre.
7. The corporate research lab and divisional labs should be performing joint applied and development projects parallel to the research projects of the university centre.

8. The firm should be hiring university graduates from the university centre.
9. The firm should join with several other firms and with government in financially supporting the university centre.
10. The firm should participate in and support several university centres sufficient to provide it with a long-term competitive edge in strategic technologies.

Technology Transfer from University to Industry in NGT

The following procedure had facilitated the appropriate kinds of cooperation for technology transfer of NGT.

1. Corporate research should strategically plan for next-generation technology, and this is best done within an industrial-university-government research consortium.
2. Corporate research should plan NGT products jointly with product development groups in the business divisions.
3. Marketing experiments should be set up and conducted jointly with corporate research and business divisions, with trial products using ideas tested in the consortium experimental prototype test-beds.
4. The CEO team should encourage long-term financial planning focused on NGT.
5. Personnel planning and personnel development are required to transition knowledge bases and skill mixes for NGT.

The consensus on vision and testing technical feasibility in a program of technology transfer between corporate research and strategic business units should pose and address the following questions:

1. What will be the boundaries of the next generation of a technology system?
2. What NGT ideas are technically demonstrable and should now be planned into product strategy, and what NGT ideas

must still be technically demonstrated and should not yet be planned into product strategy?
3. What is the pace of technical change, and when should the introduction of products based on NGT be planned?
4. What professional development and training should be planned for product development groups in order to prepare for NGT products?

Summary

Radical innovations to an existing technology occur as next-generation technology, and university-industry cooperation greatly facilitates next-generation technology innovation. Universities can research the science needed to facilitate innovation of a next-generation technology, while industries can assist in building prototypes of a NGT. Industry-university research cooperation can accelerate a nation's competitive advantage of technological discontinuities.

CHAPTER 10

University & New Ventures

Introduction

As we have been emphasizing, scientific developments and breakthroughs are important sources of technological progress. Since primary research in science is undertaken in universities, and university researchers invent new technology and even start new companies to market the technology — these are called university new ventures — how could a government assist in nurturing the process? (For example, we noted that in the history of the biotechnology industry, it was one of the co-inventors of recombinant DNA technology, Boyer, who started the first biotech company, Genentech.) Can government and industries cooperate to facilitate new venture business start-ups by university researchers? The answer is yes. First we will review a successful example of this in South Korea, the new company of Nano (Betz, Min and Shin, 2014). Then, we review a U.S. Government research program to fund research in small businesses: the SBIR program.

Nano Corporation — History

There are several types of SME businesses, namely: retail, service, and manufacturing. We focus here on the high-tech manufacturing type of SME. For these companies, the challenge lies in inventing new technology for new products and production.

In South Korea, the 'Asian Financial Crisis' of 1999 made it a difficult year; in January, the crisis threw South Korea into a sudden depression. Korean President Kim Dae Jung issued a policy to encourage university professors to start companies to provide employment for graduates since the employment growth in large firms had suddenly stopped. The government allowed professors to patent their research, and also allowed universities to provide space for professors to begin businesses, providing a grant program to fund technology development for new small businesses.

At the time, Shin Dong-woo was a professor of materials science at Gyeongsang National University. He had been doing research in nano-particle processing, and had invented a new process for producing very small titanium-oxide (TiO_2) nano-particles (20 nanometers) at a low cost. These were ten times smaller than the previous commercially available titanium-oxide particles, which measured about 200 nanometers in size. The TiO_2 nano-particles would be much more efficient as a chemical catalyst because the smaller particles could provide a larger surface to catalyse chemical reactions.

To make electrical power, many power plants burn coal or oil. The burning produces an exhaust gas containing nitrogen oxide, which is poisonous and needs to be removed. This is done by passing the exhaust fumes through a catalytic filter before passing them into the environment. The chemical reaction in the filter converts the nitrogen oxide fumes into nitrogen gas and oxygen gas. To accomplish this, a catalyst is used. A catalyst in chemistry assists a particular chemical reaction to occur faster at desirable temperatures. The catalyst in the filter is a molecule called Titanium Dioxide, TiO_2, made up of titanium (Ti) and oxygen (O). The presence of this catalyst enables the conversion of nitrous oxide (NO_x) with ammonia (NH_4OH) to nitrogen and water to occur at a lower temperature. Without the TiO_2 catalyst, the gas would have to be reheated to a high temperature for the reaction to occur. Shin applied for and received patents on his process in Korea, the United States, and China to make TiO_2 nano-particles at low cost.

Professor Shin then started a new business to produce nano-particle TiO_2 as a product, and also to employ his graduate students, majoring in Materials Engineering (who could not find other employment due to the recession). He named his new company Nano and applied to the Ministry of Environment for a grant of $200,000 to build a pilot plant in a building of 200 square metres, owned by the university. In 2000, Nano successfully completed the pilot plant, which produced a powder of TiO_2 nano-particles. Nano then acquired $5,000,000 in venture capital from investors to build a full-scale production plant.

Professor Shin contacted a fund manager who worked for the Korea Development Bank (KDB), which was at that time the largest state-owned bank in Korea. In 1999–2000, an IMF rescue fund was supplied to the Korean government, which was then transferred to governmental banks such as KDB. The manager of the fund in KDB was an invited foreign manager from Britain (who had prior accomplishments in this field). This British fund manager was looking for companies in which to invest. But demand for venture capital was diminished due to economic troubles, and there were few healthy start-up companies, resulting in a temporary surplus of venture capital looking for projects to invest in. The KDB initially gave Nano $2 million in funding. Then, KDB became a shareholder of Nano, acting as a strong and stable financial supporter. Coupled with the good reputation of the British fund manager, this helped Nano raise an additional $3 million from other Korean venture capitalists to build a full-scale plant.

In 2010, Nano received its first order for ten tons of titanium-oxide nano-particle powder from SK. SK Group is one of the Korean chaebols and the leading energy and chemicals firm. At that time, SK was looking for a partner to manufacture for them a Selective Catalytic Reduction (SCR) DeNOx catalyst that would use the TiO_2 nanoparticles.

SK had developed a recovery technology to use the spent catalyst from its own process of producing ultra-low-sulphur diesel and kerosene from the vacuum residue (VR) feedstock. They had the second largest unit in the world at that time to produce this. During the production of the diesel and kerosene, there were large amounts of impurities such as sulphur and metals in the feedstock, which fouled the catalyst, requiring a large amount of catalyst to be recharged every year. SK's new technology was able to create value from the spent catalyst, which would otherwise have had to be disposed of at a substantial cost.

In 1998, some of SK's clients were power companies that needed a TiO_2 catalytic filter to clean up their exhaust fumes, and also to periodically replace the catalyst TiO_2. But SK decided not to make the investment in producing the catalyst and filter as the market was not large enough for the conglomerate. SK decided to engage with an outside SME to manufacture it. This idea came from the R&D Center of SK Corporation. SK Corporation would form a partnership with a small firm capable of producing the DeNOx catalyst based on SK's technology. This partnership would let SK save a large portion of the required capital expenditure to build a plant to produce the catalyst. Thus, a cooperative partnership between SK (a chaebol) and Nano (a new SME) was a good way for SK to minimize the capital investment in a new commercially-viable production facility.

The first batch of TiO_2 produced by Nano was sold to SK, which needed a large amount of TiO_2 to test-produce the SCR DeNOx catalyst product, based on SK's own technology. Successful test production led to the signing of the partnership agreement between the two companies. At the same time and in the same company, the future Nano had found both a partner and a customer for their TiO_2 nano-particle product, whose technology had been invented in prior university research.

There were two products for Nano to make: the TiO_2 nano-particles, and a plastic honey-comb-shaped filter to hold them. Nano first built a facility to produce the powder, then built production equipment to construct the honeycomb frame (called SCR), which held the catalytic titanium oxide nano-particles. In 2003, supplying both the catalytic powder and honeycomb SCR filter, Nano fulfilled its order to SK.

However, in 2002, Nano had also supplied its honeycomb filter and catalyst to the Bundang Complex Power Plant (before Nano and SK signed a partnership agreement in 2003). So Nano began with not just SK as its sole customer, but also with another client, a power plant. This was the first step toward Nano avoiding having only one large company as a single customer.

Bundang is a public company, which supplies heat and power to the Bundang municipality located in the south of Seoul. It is a small power plant and does not require a large amount of DeNOx catalyst. Nano was able to supply this company with its own product produced by the pilot facilities. As this was before Nano and SK signed the partnership, Nano had its own "recipe" to manufacture DeNOx catalyst without using the powder and the recipe which used the spent catalyst from the heavy-oil desulfurization unit. Thus, there was no problem between SK and Nano regarding intellectual property.

At the time just after the Bundang project, Nano thought it needed more than just catalyst-making technology to become a serious contender in the DeNOx catalyst market. It needed the appropriate production capacity to ensure quantity, quality, and low cost. If a vendor wanted to supply DeNOx catalyst to KEPCO (Korea Electric Power Company), the single largest customer in Korea, it had to meet strict eligibility requirements. Only foreign companies with a long history of manufacturing and marketing were qualified to bid for the DeNOx catalyst contract, due to the stringent requirements. Still, Nano was determined to meet them and become a DeNOx catalyst supplier.

In 2003, Nano supplied its first export of the honeycomb SCR catalyst to EnBW in Germany. This effort required overcoming huge problems in producing the new product. It was delivered on time by flying the very heavy product to Germany. There, the TiO_2 nano-particle powder was applied as a catalyst substrate in the stream of exhaust gases from coal combustion. The honeycomb shape of the SCR catalyst was instrumental in this process.

Nano had developed two kinds of production processes: to produce the nano-particle powder and to extrude the honeycomb. During 2004, Nano continued to improve both their production processes of producing Titanium Oxide and extruding the honeycomb filter. These efforts were helped by collaborations between Nano and a couple of other Korean SMEs. Another research grant of $2.5 million from the Korean government enabled Nano to fund research to continue improving its production technology. Cooperation between government and university new-ventures and a large chaebol firm had assisted Nano to continue to grow into a substantial business.

In 2005, SK-Nano supplied a honeycomb SCR to a Korean Coal Fired Power Plant in Boryong. In 2007, Nano supplied a honeycomb SCR to a Heavy Oil Power Plant in Ulsan, independent from SK. It also supplied honeycomb SCRs to Diesel Power Plants on Jeju Island.

In addition to these new customers, Nano expanded its business by adding services along with their products. In 2008, Nano began providing catalyst maintenance services to replace the catalysts in SCR honeycombs, with Samchunpo Power Plant being their first customer.

To help finance this service expansion, new research and development (R&D) efforts had been made and initiated in 2002, partly with government funding. Nano developed the equipment for all its production needs (equipment, extruders, dryers and other critical units) in order to reduce production cost and control quality.

As business expanded, Nano also needed to expand production capacity. In 2009, Nano built another facility in Sangju, this time 50,000 square meters large, for producing SCR catalysts using Nano's developed production equipment.

Nano had begun with research from the university and then continued research to improve production capabilities, remaining high-tech in R&D. In 2009, Nano hosted a 3rd International Conference for SCR maintenance technology, broadening its exposure and contacts with the technology community. It also received another research grant from the Korean government to develop a SCR catalyst for marine applications.

In 2010, Nano's success at business and high tech was recognized as they received a presidential award for top technology. It was nominated by the Korea Credit Guarantee Fund to be one of Korea's top five green technology companies. Nano hosted a 4th International Conference for SCR Technology in Busan. It was also appointed as candidate of hidden champion enterprises by Korea Export Import bank.

In 2012, Nano built a new R&D center and a second production plant at Sangju, continuing its R&D efforts and generating new SCR patents. It also hosted the 5th International Conference for SCR Technology in Busan in June, and opened an office in Shanghai, China: NANO CHINA CO. In the same year, having anticipated a growing market in applying SCR technology to diesel-powered commercial ships, Nano aimed to become a leading manufacturer in the emerging SCR market for marine applications. Korea was one of the dominant ship builders in the global market, which helped Nano's position in this newly opened market.

In 2013, a third large investment increased the capacity of honeycomb production up to 6,000 cubic meters by June, and a new plate type catalyst increased annual capacity to 10,000 cubic meters by the end of 2013. A 6th International Conference for SCR Technology was held in Busan in September 2013. Nano's revenue rose from $50 million in 2013 to $100 million in 2014 as a result of new markets formed in China from 2013. In 2014, 90% of Nano's revenue was from export to foreign markets.

In this case of Nano, we see that many cooperative relationships encouraged and assisted its establishment and growth. Government policy encouraged the university to allow its professor space to build a prototype and production facility; and a grant from a government-assisted international bank provided the professor with a loan to build the first production facility. Cooperative strategy with a large chaebol, SK, provided Nano with its first customer, and assisted them in finding other early customers. A kind of 'middlestadt' orientation on the part of the founding professor-president of Nano focused his ambition to provide employment for his graduates and to assist Nano's employees to further education.

Innovation Theory: New Venture Milestones

The business literature on new ventures has generalized a pattern of accomplishments (milestones), and new business ventures must meet these for survival and success. We briefly review new venture theory, and examine the milestones in the case of Nano. University-based new ventures must succeed commercially, just like any commercial venture.

The theory of new ventures consists of several milestones, which new ventures have to meet:

- *Acquisition of start-up capital,*
- *Development of new product or service,*
- *Establishment of production capabilities,*
- *Sales growth and production expansion,*
- *Meeting competitive challenges,*
- *Product improvement, production improvement and product diversification,*
- *Organizational and management development,*
- *Capital liquidity.*

(Betz, 2015)

We next see that Nano did meet all these milestone challenges, as the role of a faculty member grew to encompass the role of an entrepreneur. This is a necessary personnel transition in any university-based new venture — from faculty to business leader.

Acquisition of Nano's start-up capital was, as we recall, the initial KDB fund decided by the British fund manager, and the funding by this highly reputed British fund manager readily attracted up to \$5 million of other venture capital within a month. With that money, facilities were built for the production of TiO_2 (Nano's main product at the initial stage), and to manufacture an SCR DeNOx catalyst. This was a fully integrated unit, although of pilot plant scale. The first batch of DeNOx catalyst supplied to Bundang Cogeneration Plant was produced using this pilot-scale unit, with the university providing the building for the pilot plant facility.

For the development of a new product, the TiO_2 and DeNOx catalyst of Nano's own manufacturing recipe was developed by employees using the facilities. Most of them were Professor Shin's graduate students. For the development of processing machines and testing and evaluation equipment, NANO took advantage of the \$2.5 million four-year governmental R&D fund granted in 2002. During this initial period, Nano followed two developmental paths. One was the production of catalyst for a captive user, SK (using the hardware provided by SK and Nano). Second was the development of improved cost-effective technologies by Nano itself, from processing, monitoring, and quality-control to production machinery.

For the establishment of production capabilities, we recall that Nano had the capability to produce the DeNOx catalyst on its own, even before the SK-Nano partnership was formed. Nano simply had to produce the DeNOx catalyst based on the SK recipe, which was a key element of the collaboration agreement.

As for sales growth, even with the collaborative effort with SK Corporation, the rate of sales was initially slow and disappointing. This was due to the KEPCO's stringent requirements. The breakthrough for Nano occurred when ENBW, the third largest power company in the European Union, decided to buy the SK-Nano product. The decision was made after a series of bench-scale tests at ENBW and an all-out marketing effort by SK Corporation. ENBW, after being fully convinced by SK's impressive R&D capabilities, and with a heavy price discount offer from the Korean partnership, finally decided to buy roughly one million dollars of product. In return, the companies delivered quality products on time. Towards the last stage of the product delivery, SK Corporation paid a high airfreight bill to deliver the DeNOx catalyst for the power plant start-up. Due to ENBW sales and satisfactory performance records at the German power plant, SK and Nano became the sole domestic company eligible to sell DeNOx catalyst to KEPCO — the single largest Korean user of the product. Soon, the market share of the SK-Nano DeNOx catalyst soared to over 60% in the domestic market. SK (a large conglomerate) helped Nano (a new SME) to find a significant first customer other than SK.

For production expansion, Nano needed to build new plants twice. Its choice for its first DeNOx manufacturing plant was to locate it in the suburban area of the southernmost city called Jinju (which is still known as the city of school teachers because of the high ratio of teachers to population). All the professional employees of Nano, as well as its CEO, were either graduates from or a professor at the National University of that small city of education. But with continuing sales growth, both in the domestic and the foreign markets, there arose a need to expand Nano's original 2,000 cubic meters/year capacities to a higher level; and this became a crucial issue in the partnership with SK. Unfortunately, SK's interest in the partnership waned. The manger of SK Corporation, who first encouraged SK's participation, had moved to another position within the SK Group CTO office. His successor was not interested in the strategic alliances with SME companies. One issue at the time was whether SK might be interested in helping to finance production expansion in Nano. SK was not interested then. That partnership, which had lasted for over 8 years, ended. Nano was confident in its independence since it had developed all subsequent production technologies and equipment by itself. Nano needed to expand its capacity, so it began looking in areas outside Jinju for potential sites for a new plant that would be quadruple the size of the first plant in

Jinju. The city of Sangju, located in another province, made an extremely favourable offer to Nano, expecting that Nano would bring jobs and tax revenues to that city. Two banks, IBK and KDB (which are essentially owned by the government), were competing against each other to become the fund provider to Nano. This is partly due to the policy of the Korean government, which wielded great influence over those two banks. President Lee of the Republic of Korea was at that time pushing for strong SME policy in order to overcome the world economic crisis of 2008–2009. Everything in the government policy on SMEs at that time continued to be in Nano's favour. This was not planned, but pure luck.

Meeting competitive challenges required that Nano face competitive challenges against both domestic and foreign competitors. Since the financial crisis, which had hit Korea in 1998, several Korean companies had attempted to establish themselves as viable players in the domestic market. None of them were able to match Nano's capabilities in terms of product quality, competitiveness in cost and production capacity. A couple of much smaller companies were competing against Nano, but only in the very small market of cogeneration and garbage incineration plants. Hence, there were no serious domestic competitors for Nano in the domestic market. In the overseas market, the situation was different in that a number of companies with sizable production capacity and cost competitiveness were fighting for the market share. There was a US-Japanese joint venture called Cormetec, and most of the serious players in the DeNOx catalyst market were European companies. But by 2013, Nano was almost the sole player in the DeNOx catalyst market that had expanded its production capacity (which might have a lot to do with the depressed economies of both European and Japan in 2013). Also in 2013, the marine application sector emerged as a prospective market for the DeNOx catalyst. Nano has been aggressively pushing to become the leader in this emerging area, whereas it had been a follower in the DeNOx catalyst market. The fact that Korea was one of the leading countries in the global ship-building industry may have put Nano in a favourable position for the new DeNOx market for marine application. One German company, which designs the SCR DeNOx system for the big ships propelled by large-size diesel engines, has formed a partnership with Nano in that emerging marine DeNOx market. This indicates the excellent reputation Nano had achieved in the international market.

For product improvement and production improvement, Nano had steadily improved its product quality to such a point that it could satisfy the quality expectations from the more sophisticated European electric power generators, as well as the customers in Asian companies (including Japanese ones). One of Nano's accomplishments was in production improvement. All the equipment and facilities associated with DeNOx catalyst manufacturing are manufactured and supplied by Korean SME companies that have close working relationships with Nano. R&D funding from the Korean government played a critical role in letting Nano and domestic SME partner companies come up with a complete set of technological capabilities to fabricate essentially all the equipment and facilities used in the process of DeNOx catalyst manufacturing. This put Nano in a position to compete against any foreign catalyst-manufacturing companies.

For product diversification, Nano developed a regeneration technology which allowed waste catalyst to be used again. This technology competed with fresh catalyst modules and was accepted by most of the Korean power plants because it halved the cost. It was a lucrative business in 2010–11 before many of the market followers duplicated a similar technology. In addition, Nano steadily developed the catalyst for marine application, which required high resistance to vibration, and have become a global top supplier for the marine market since 2012. Nano has also developed the plate type SCR catalyst, which occupied 50% of the domestic and world SCR market, with only three other companies in Japan and Europe competing to manufacture it. By the end of 2013, a new production line for the plate type SCR catalyst with a capacity of 10,000 cubic meters was completed. At the same time, the production line for the honeycomb catalyst for marine application had increased to 6,000 cubic meters in 2013.

Innovation Helixes

Earlier we reviewed a temporal picture of innovation structures that conceptualized them as being helical over time. The model emphasizes the interactions between government, university, and industry. We recall that in 2009, E.G. Carayannis and D. F. J. Campbell extended this triple-helix model to a quadruple-helix one by adding another interaction to the innovation process. The last helix represented the civic public in the forms of media and culture. (Carayannis and Campbell, 2009, 2010) In 2012, they expanded on the elements of the fourth helix of civil society and culture, as shown in Figure 10.1. (Carayannis and Campbell, 2012)

This depiction of the four helical forms in the processes of innovation emphasize that these institutional processes occur in the three research sectors of government, university, and industry, and are facilitated by a civil society and culture. It emphasizes that they are all cyclic in two ways. The knowledge processes in each sector are cyclic, and so is the way knowledge is passed between sectors, each focused upon 'funding', 'science', 'technology' and 'culture' respectively. In this case study, we have examined the interconnecting processes for innovation as experienced by a new SME in Korea. Nano Corporation was a new venture, beginning in South Korea in 1999. It was created to make use of university research, and was assisted by government grants and industrial cooperation to emerge later as the market leader in SCR catalysts.

Also we recall that the German 'Mittelstand' is a particularly interesting quadruple helix innovation process in SMEs. South Korea became interested in the Mittelstand concept because the German SME's strong contribution to the German export

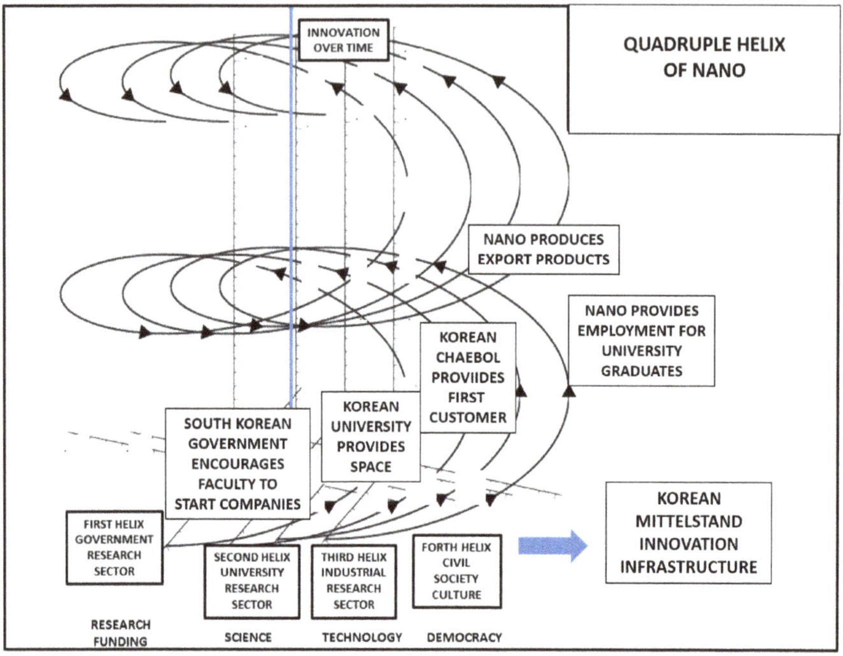

Figure 10.1. Quadruple helix depiction of nano innovations.

economy had been recognized as significant. In this case study, we have seen that a 'Mittelstand culture' was incorporated into Nano's strategy for becoming a major exporter of new high-tech products in the global market.

At the time of its founding, Nano was not technically a 'family business'. But it was founded with a kind of 'intellectual family' in mind. Professor Shin specifically began the company to provide employment for his materials engineering graduate students, and his former graduate students still form the upper management level of the new company. Nano focused on global market dominance in the SCR market for coal-fired electrical power plants, and next, for diesel-powered commercial ships. They also had continuing R&D efforts to improve operational effectiveness (in its processes of producing nano-particle catalytic powders and extruding honey-comb filters).

Although the Korean education system does not have a focus on vocational training like the German education system, and the equivalent of the German Fraunhofer Institutes have not yet been institutionalized in the Korean national innovation system, Nano has had a unique attitude towards the educational system. Nano consistently encouraged employees to attend Masters and Ph.D courses at the university, where Professor Shin continued as a faculty member. Their theses for advanced degrees were relevant to the product of Nano. Their experiments were carried out at the Nano R&D centre. The time and tuition fee for advanced degrees were fully supported by Nano as company policy. As a result of this part-time degree program, three doctorates were awarded to Nano employees. They began working in Nano as heads of the production, R&D and raw powder materials divisions. There were ten other employees with Masters Degrees in various majors, including management. This part-time degree program is still operating with one Ph.D. candidate and two masters candidates in diversified fields.

Illustration: US Small Business Innovation Research Program

In the United States, the Fraunhofer model was never copied. In 1982, the U.S. established a new program for the government funding of small businesses called Small Business Innovation Research (SBIR). All government research-funding agencies were required to spend 2.5% of their research budgets on research grants to small businesses. In 2017, the percentage of funding was subsequently raised to 3.2%. Federal agencies participating in the program include the Department of Defence, the National Institutes of Health, the National Science Foundation, and the Department of Agriculture.

The SBIR program published: "The mission of the SBIR program is to support scientific excellence and technological innovation through the investment of Federal research funds in critical American priorities to build a strong national economy. The program's goals are four-fold:

- Stimulate technological innovation.
- Meet Federal research and development needs.
- Foster and encourage participation in innovation and entrepreneurship for socially and economically disadvantaged persons.
- Increase private-sector commercialization of innovations derived from Federal research and development funding." (SBIR-Participating Agencies, 2017)

The SBIR program had federal agencies make awards in two phases (of a three-phase plan for innovation). Government SBIR awards were only for phases I and II:

- "**Phase I**, the start-up phase, makes awards of "up to $150,000 for approximately 6 months support [for] exploration of the technical merit, or feasibility of an idea or technology."
- "**Phase II** awards grants of "up to $1 million, for as many as 2 years," in order to facilitate the expansion of the results of Phase I. Research and development work is performed and the developer evaluates the potential for commercialization." (SBIR, n.d.)
- **Phase III** is the phase in which commercialization of the research-based innovation begins. This commercialization is to be wholly funded by private sector funds. DOD published: "Phase III commercialization work derives from, extends, or logically concludes efforts performed under prior SBIR/STTR funding agreements. Small businesses are expected to obtain additional funds from the private sector, and/or non-SBIR/non-STTR government sources to develop Phase II prototypes into viable products for sale in the marketplace." (U.S. Department of Defence, n.d.)

For example, in 2016, the US Department of Defence spent about $1 billion dollars in SBIR grants. Half of the awards went to firms employing less than 25 people, and one third to firms with fewer than 10 employees; a fifth of the awards went to businesses owned by women or minorities. Annually about one quarter of the grants went to new grantees and three-quarters went to businesses which had earlier won a DOD SBIR grant. DOD published: "Historically, about 15% of SBIR and STTR proposals are awarded a Phase I contract; approximately 50 % of Phase I projects subsequently are awarded a Phase II contract." (U.S. Department of Defence, n.d.)

In 2012, the U.S. Congress reauthorized the SBIR program for five more years. At that time, the program director in the NSF SBIR Program was Roland Tibbets. In testimony for the congressional reauthorization, Tibbets wrote: "SBIR addresses a paradox at the heart of innovation funding: capital is always short until the test results are in. At the idea stage, and even the early development stage, the risks are too great for all but a few investors. But innovations can't get beyond that stage without funding. The business models of most investors focus on generating many sales to many customers." (Tibbets, 2009)

Tibbets also emphasized that when the government is a potential customer for a small business innovation, there is also an early funding problem: "There is another paradox, too. The federal government has R&D needs that, for a variety of reasons, will never interest private sector investors. When the government is the only buyer, and buys on a one-time or very occasional basis, investors get skittish. Large government contractors typically aren't interested in such R&D, either. The amounts involved are too small, and most large contractors don't have early-stage R&D capabilities anyway. So needed innovations in fields like defence, space exploration and homeland security may not occur." (Tibbets, 2009)

Tibbets also wrote that the SBIR program was designed to address both these early-stage innovation funding problems: "SBIR was designed specifically to solve both of these paradoxes:

- First, it provides a transparent, competitive and reliable source of early-stage funding for R&D, based entirely on scientific merit. Today, SBIR is the nation's largest source of such funding.

- Second, it allows the government itself to obtain needed R&D that the private sector could not otherwise provide... SBIR's success, as recently documented by the major National Research Council / National Academy of Sciences study, is rooted in a number of the program's characteristics.

Drawing on small business scientific talent, SBIR draws on the six million scientists and engineers that are now employed by small firms. That compares to the five million employed by medium-sized and large firms. In fact, small business employs more scientists and engineers than large business, universities, federal labs, or non-profit organizations... According to the National Science Foundation's annual Science Indicators report (2012), large firms receive 50.3% of federal R&D, universities receive 35.3%, non-profits 10% and small businesses just 4.3%. SBIR accounts for over half of that 4.3%... For small companies, SBIR remains the only game in town, just as it was in 1983, when it began." (Tibbets, 2009)

Summary

This case offers evidence that the four strands of helical innovation interactions are all important.

- Firstly, the university, industrial, and governmental research sectors interact in innovating new high-tech SME businesses.
- Secondly, government research funding agencies should have programs to sponsor research grants from SMEs to improve technical processes.
- Thirdly, strategic technology alliances between SME suppliers and large conglomerates are important to grow a new technology market, which may be too small for the conglomerate, but very significant for the SME.
- Finally, the proper civil culture and proper government policies are important to encourage university professors

to establish new high-tech businesses from their university-based research.

SBIR programs provide funding to individuals and small businesses for one year of research, and also a possible continuation of two more years of R&D — for developing a new product or service. The SBIR government agencies hope some of these products/services will develop into new businesses, funded by venture capital sources. Without an active venture capital sector that is willing to fund a new start-up, the SBIR plan is incomplete. An SBIR program sustained over time will certainly create a small sector of contract research organizations. But whether or not such contract-research firms evolve and successfully become small commercial businesses — this wholly depends upon private capital funds (venture capital) and not on the government. *SBIR programs require a cooperative research partnership between a government and venture capital.*

Chapter 11

Technology Discontinuities & Creative Destruction

Introduction

We have emphasized how research cooperation between universities, industries and governments can facilitate dramatic innovation which may lead to the invention of a next-generation technology. This is a discontinuity in technological progress — it creates not a small increase in technical performance, but a great leap forward! Economically, this is important because technology discontinuities create economic discontinuities, both destroying existing industries and creating new industries, resulting in great economic opportunity.

Earlier, we noted that the concept of 'innovation' consisted of at least two parts: 'invention' and 'commercialization'. These occur in the early part of innovation; but in the long run, there is a third part — 'market-capture'. Innovation requires creating a technical idea, making products in which the technical idea is embedded, and then beating the competition with a product that has better technology. We have focused on the invention and commercialization of innovation in cooperative research partnerships in the previous chapters. In this chapter, we will look at market capture, the third and longest-term aspect of innovation.

In the twentieth century, Americans excelled at the early parts of innovation, 'invention' and 'commercialization'. But after the Second World War, American firms did not do as well in 'long-term market capture'. American firms invested in the invention and commercialization of new high-tech products, but not in the long-term capture of markets. For example, American firms invented and commercialized the transistor (but only in business and military products). Next, Japanese firms commercialized transistors in consumer electronics, capturing the world's consumer electronic markets. Then, American firms innovated Integrated Circuit Chips (IC chips), but soon lost the IC memory chip market to Japanese firms. How did this happen?

Over the longer term of innovation, cooperative innovation involves not only science and technology, but also capital. Yet in the last decades of the twentieth century, management in American industry lost interest in investing for the 'long term' and maintaining market dominance. Instead, American capital (as traded on Wall Street) began focusing on short-term profits (and equity price). American capital markets had lost their staying power.

Next-Generation Technologies and Creative Destruction

To understand the impact of a 'short-term' focus on innovation, we will review how technology discontinuities impact industrial dynamics by altering industrial competition and structure. The economist Peter Schumpeter called dramatic changes of industrial dynamics 'creative destruction'.

The creative destruction of economic systems characterized economic history in the twentieth century. One case of 'creative destruction' was the changing positions of countries in the rankings of successful IC chip producers. From 1950 to 1980, the United States was a success story for innovation and competitiveness, innovating and becoming highly competitive in the computer and electronics industries (particularly transistors and semiconducting integrated circuits). But from 1980 to 1989, Japan became the star of the next success story, and was the most competitive country, but not necessarily for innovation. The U.S. still led in the innovation of computers, the Internet, and biotechnology, but no longer led in long-term commercial competitiveness. In the 1980s, Japan led the world in commercial competitiveness. By the 2000s, Samsung and South Korea gained the competitive edge in the memory chip industry.

We briefly review this history of the electronics industry to see the importance of the connections between research and technology and capital in determining 'market-capture'.

History: Samsung's Creative Destruction in the Memory Chip Industry — 2000

In 1983, there were two prior historical events (Moore's law and Japan's creative destruction in memory chips, both of which we will soon recount) that set the stage for the ambition of a South Korean firm, Samsung, to become the world leader in DRAM memory chip production and capture the IC memory chip market.

Tony Michell (2012) wrote: "Why did Samsung Electronics... continue to invest money in (electronic) components? The answer in this strategy lay at the heart of what it meant to be Korean in the... early 1980s. Despite escaping from Japanese colonization in 1945, Korean industry remained heavily dependent on Japanese know-how and even more on Japanese components... Unless they could penetrate into the component business, Korean industry would forever be dependent on the Japanese, who were now dominating the electronics industry... Chairman Lee in February 1983 announced that Samsung would make full inroads into the semiconductor industry."

It took two years, from 1983 to 1985, before Samsung produced its 64K DRAM chip. This was four years after their Japanese competitors had produced their 64K chip in 1981. There was a lot of catching up to do before Korea could compete with Japan in IC chips. But they accomplished all this, as is summarized in Figure 11.1.

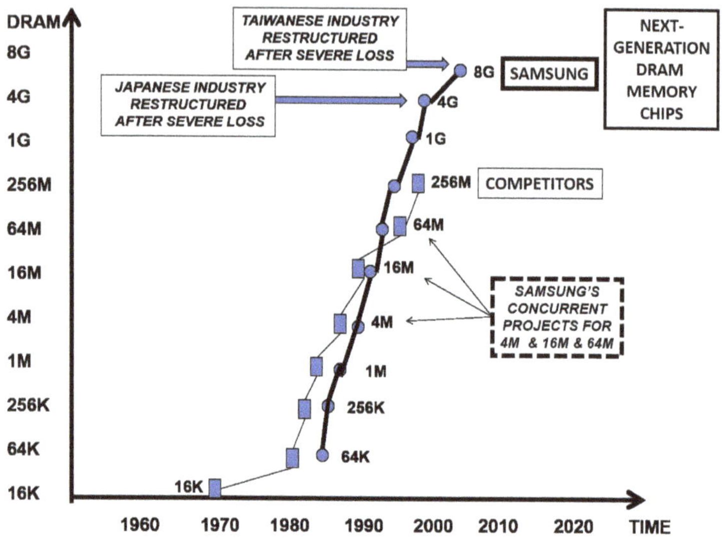

Figure 11.1. Progress in DRAM IC chips from 1970 to 2010, with creative destructions

In 1985, Samsung's 64K DRAM memory chip was late into the world market. The global IC chip technology race was following the plan of Moore's law (which we will soon discuss) and had already innovated the 256K chip; the 1M chip was being developed. Samsung had to catch up with international competitors in the technology race before it could make substantial profits on its investment. In 1981, the U.S. chipmakers had been forced to drop out of the race, at the 64K technology discontinuity (which we will soon review). No longer were they making enough sales and revenue to be able to afford investing in the next-generation-technology 256K chip. Finally in 1992, Samsung innovated the 64 M DRAM chip just ahead of competitors and then continued to lead in time. By 1995, Samsung was dominating the DRAM market with its 4G memory chip, and Japanese firms left the market. In 2000, Taiwanese firms stopped producing DRAM chips when Samsung introduced the

8G chip. Samsung had executed 'creative destruction' on both the Japanese and Taiwanese memory chip makers.

Why was Samsung able to enter the race late and stay in it and eventually win it? It was because the Samsung group had both the determination and capital to win the race. The Samsung Group was a Korean chaebol. It could finance the continuing development of memory chips, not from selling memory chips, but from selling other products through its chaebol member companies. The structure of capital in Korean chaebols was important to the Korean firms' capability to continue investing in manufacturing development for the long duration of the chip race — corresponding to the strategy of Moore's law.

Within a chaebol capital structure, Samsung Electronics began development of the VLSI stage of the 64K DRAM chip in May 1983. It contracted the U.S. firm Micron Technology for the chip design, and a Japanese firm for the production technology. By November 1983, six months later, Samsung succeeded in the development of the 64K DRAM chip. But it was still two years behind Japanese producers. So in September 1983, just as the Korean group was finishing the 64K chip, a Samsung group in America began designing a 246K DRAM chip. Samsung was concurrently developing not one, but two generations of DRAM chips in 1983. This was Samsung's strategy to 'catch up and leap forward'.

In March 1984, Samsung had established the second group in Korea to work on the 246K DRAM chip. Yet in 1984, the 256K DRAM chip already was being commercialized (produced and sold) by Japanese firms. Samsung would not yet 'catch up' in technology at the 256K DRAM stage. In 1984, Samsung began a development project for the 1M DRAM chip (which it finished in 1986). Still even at the 1M chip, Samsung lagged behind Japanese competitors by two years. Samsung still could not make substantial profits until it caught up. In late 1986, Samsung launched a concurrent project for the 4M DRAM chip (accomplishing it in 1988). Then Samsung was only a half year behind competitors. So in 1987, Samsung started another concurrent project to develop the 16M chip, which it finished in 1989, at the same time as its competitors. Now, Samsung could export DRAM chips of 16M capacity, and make substantial profits after five years of investment. Samsung had caught up with competition by concurrently developing three generations of DRAM chips: 1M, 4M, and 16M chips in parallel.

But catching up was not sufficient to 'conquer' the competition. Samsung wanted to and needed to 'leap ahead'. In 1989 (instead of just producing 16M chips to recover investment), Samsung started a concurrent project for the 64M DRAM chip. It produced this in 1992; now, it had a small lead on its competition! Samsung was finally earning export profits from memory chip sales and forged ahead to extend its lead. A 256M chip project was begun concurrently in 1991 (while the 16M chip was still being developed). The 256M chip successfully went commercial in 1994, giving Samsung a one-year lead ahead of its competitors. By then, Samsung was dominating the memory chip market. This had been Byung-Chull Lee's ambition in 1983, and his son, Kun-Hee Lee, achieved memory chip dominance in 1994, a decade later.

Samsung pressed on, continuing the race. In 1993, as the 256M chip was still being developed, Samsung had a second team starting the 1G chip. By 1996, Samsung had innovated the 1G chip ahead of competitors by one and a half years. Samsung had begun the 4G chip development concurrently in 1993 and commercialized it in 2001.

Millions of dollars had been used to establish production capacities for all those generations of DRAM chips (16M, 256M, 1G, 4G). In 2010, Samsung received enough cash flow to earn back all its investment and much more. The lagging competitors didn't earn profits. In 2000, Japanese firms (which had seized the memory chip business from the U.S. twenty years earlier) quit the memory chip business after severe financial losses. Korea had conquered in the Moore's Law race of technology discontinuities. And in doing so, Samsung created a Schumpeterian creative-destruction of the Japanese DRAM chip industry.

Samsung went on in the 2000s to innovate the 8G, 16G, and 32G chips. Then the Taiwanese chip competitors dropped out. By 2005, Samsung had dominated the DRAM memory business by 32% and the NAND flash memory business by 51%. In 2010, Samsung earned 10 billion dollars profit alone from its semiconductors. Each year, Samsung earned several times its investment to capture the market.

Industrial Dynamics

The invention of electronics has had a huge impact on the economy. Electronics enables systems and devices to be smaller and faster and has paved the way for many inventions, such as mobile devices, Internet of Things, and cloud computing, that we have come to know of today. In 1942, to explain the connection between innovation and economic structure, Joseph Schumpeter, a professor at Harvard University, coined the term 'creative destruction'. Radical innovations create a new structure in an industry but also destroy a previous industrial structure, resulting in technological and economic discontinuities (Schumpeter, 1942).

In 1981, David Ford and Chris Ryan (1981) proposed that each industry be represented by a chart of market volume over time. This would reflect the underlying maturation of the core technologies of its product; they called it an 'industrial life cycle'. William Abernathy and K. B. Clark in 1985 described how in the early days of this pattern, the product innovations peaked and then were followed by production innovations (Abernathy and Clark, 1985). In 1988, the author combined the industrial life cycle with the product technology S-curve and the numbers of industrial competitors to show their relationships (see Figure 11.2) (Betz, 2010).

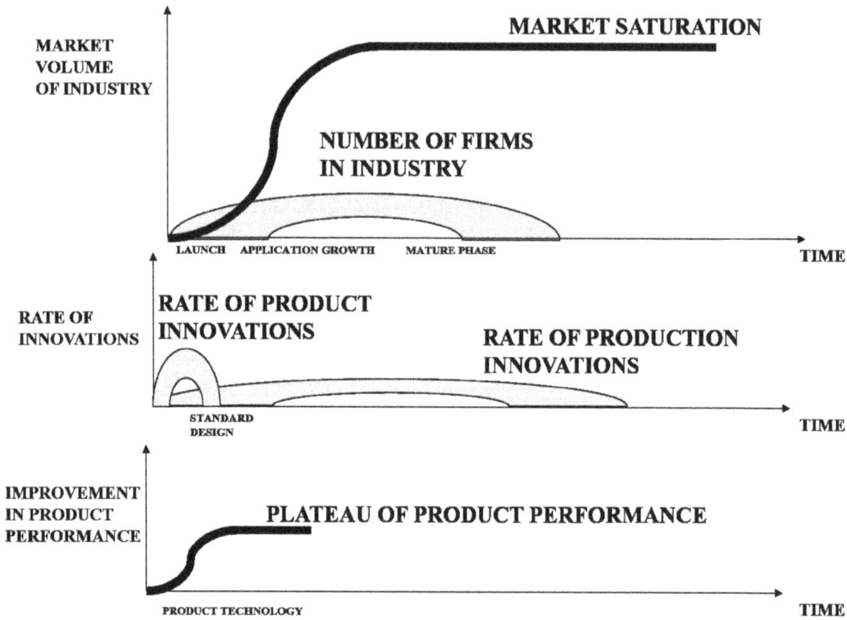

Figure 11.2. Dynamics of new industries based on technological innovation.

Based on new radical technology, new products create new industries (e.g. the invention of the stored program electronic computer created the computer industry in 1950). In the launch phase, many new product technology innovations spur the growth of the new industry. But after a product design standard is achieved, the rate of product innovations decline. Instead, the rate of production innovations (to produce products, reduce costs, and improve quality) increases. As technical innovation both in products and production slows, the industry enters a mature industrial phase with commodity-like products. All the products in the industry are similar in performance and cost, with differences occurring principally in product design between competitors. The number of firms in the new industry increases rapidly in the launch phase, peaks in the application growth phase then declines in the mature industry phase.

This pattern depicts the historical pattern of new high-tech industries, but only when there are no further technology discontinuities in the industry. However, the electronics industries showed repeated discontinuities in generations of new technology — from vacuum tubes to transistors, to integrated circuit chips, to generations of IC chip products. The history of the IC semiconductor chip industry is one of several technology discontinuities that result in the 'next generation of technologies' (NGTs). The usual industrial life-cycle pattern has not been historically accurate for the electronics industry. We will return to this Schumpeterian discontinuity theory after reviewing more of the history of electronics technology.

Creative Destructions in the Electronics Industry

Until the innovation of integrated circuit (IC) chips, there had been two prior radical innovations in electronics, the electron vacuum tube and then the transistor. In 1906 electronics began as an industry with the invention of the electron vacuum tube. Electrons were discovered in 1894 by J. J. Thomson, a British physicist. In 1906, Lee de Forest invented the electron vacuum tube, an electronic device to control the flow of electrons in a circuit. Electrons were sent through a vacuum in a glass tube from a negatively charged cathode plate to a positively charged anode plate. The intensity of the flow of electrons could be modified by a 'grid' (a mesh of wires between the cathode and anode) to which a controlling voltage could be applied. In this way, an electron vacuum tube controls electronic signals within an electronic circuit. This was the beginning of modern electronics. The electron vacuum tube made possible many subsequent electronic inventions, including radio and television.

But the invention had technical limitations. Tubes were relatively big and used a lot of electricity to heat the anode to boil off electrons. By the middle of the twentieth century, an electronic device to control electricity was needed. The research laboratory which took on the task to replace the electron vacuum tube was Bell Laboratories in the U.S.

When the electron vacuum tube was invented, AT&T became the centralized U.S. telephone company to provide long-distance telephone service to the nation. Then engineers and scientists in AT&T's Bell Labs leapt upon the invention of the electron vacuum tube and innovated the first mass production of electron vacuum tubes. AT&T used them to amplify phone conversations. By the 1930s, only twenty years after the innovation of the electron vacuum tube, some of Bell Lab's scientists were thinking of replacing the electron vacuum tube with a better device, but did not know how or what to use. This was a challenge that Mervin Kelly of Bell Labs took up in 1936. Kelly had been appointed director of research at Bell Labs, and he established a research group to engineer a replacement for the electron vacuum tube (Wolff, 1983).

Kelly was trained as a physicist, having followed the exciting developments in physics of the new quantum mechanics of the 1920s. Kelly saw a technological opportunity in quantum mechanics as it provided a theory to explain the behaviour of electrons in materials. Kelly envisioned that the new theory, when applied to certain semi-conducting materials like germanium and silicon, might provide a way to invent something to replace the electron vacuum tube. In 1936, Kelly created a research program in Bell labs in solid-state physics, to explore the use of the materials of germanium and silicon for electronic control devices.

One of the new hires Kelly made for his new program was William Shockley, who had just received his Ph.D. in solid-state physics from MIT. Michael Wolff wrote, quoting Shockley: Kelly gave me (Shockley) an eloquent pep talk... He pointed out that relays in telephone exchanges caused problems... He felt electronics should contribute... Dr. Kelly's discussion left me continually alert for possible applications of solid-state effects in telephone switching problems." (Wolff, 1983) Kelly's support and Shockley's' alertness paid off — but only after World War II.

During the War, Kelly was responsible for 1200 military research programs performed at Bell Labs, and Shockley spent the war at the Pentagon, working on military problems.

After the war, Kelly re-established the solid-state physics research program, and invited Shockley back to Bell. He also appointed James Fisk to assistant director of physical research at Bell and told him to make a major and broad research effort in solid-state physics and theory. Fisk set up three research groups: physical electronics, electron dynamics, and solid-state physics. William Shockley and Stanley Morgan led the solid-state physics group. They hired John Bardeen and Walter Brattain, new solid-state physicists, into the group.

Shockley's group learned how to alter the conductivity of germanium by adding other atoms (doping) to the germanium crystal during its growth. By doping (adding extra atoms) into a germanium crystal, they learned that they could change the quantity of electron flow in the crystal. Next, Shockley proposed an artifact (transistor) to control the current in doped-germanium crystals by applying an external electric field perpendicular to the direction of current in the crystal. This worked similarly to how a voltage applied to the grid of an electron vacuum tube controlled the flow of electrons in the tube. The reason Shockley thought this was possible for germanium was because he had used quantum theory to calculate the expected effect of a controlling field on the electron flow. His calculations showed that it should be large enough to provide amplification. Amplification meant that the electron flow through germanium would closely follow the increases and decreases of the controlling field, thereby mimicking its form but at a larger amplitude. Shockley made a sample, but it did not work as intended. Shockley asked Bardeen, a recent graduate in physics from Columbia, to check his calculation. Bardeen did and suggested that perhaps electrons were being trapped at the surface of the crystal, which would prevent them seeing the calculated effect of amplification. The science of 'quantum mechanics' enabled them to calculate that they were on the right track to accomplishing the desired technology of 'amplification' of an electronic signal (using the doped germanium crystal), but there was still a technical problem.

Next, Brattain performed an experiment on the artifact, which confirmed Bardeen's explanation. They were on the right track, but the structure of the artifact needed to be modified and the team had to try other configurations. Finally, one configuration worked, but the resulting amplification was still too small. The group continued to research alternative configurations. Another configuration placed two closely spaced gold strips on the surface of a dope germanium crystal and an electrical connection on its base. Relative to that base, one gold strip, the emitter, had a positive voltage, and the other, the collector, had a negative voltage. They applied a signal between the emitter and the base, finally achieving a sufficiently large amplification in the voltage between the collector and the base. They called this a 'transistor'. It could do everything the old electron tube could and would, and eventually replaced it. *The transistor was 'next-generation-technology', replacing the electron-vacuum tube.*

AT&T acquired a basic patent for the transistor and made transistors for its own use. AT&T licensed others to make them for other uses. At that time, AT&T was a legal telephone monopoly in the U.S. and was legally prevented from entering other businesses. The invention of the electron vacuum tube saw the dawn of the first age of electronics. The transistor revolutionised the second. Shockley, Brattain and Bardeen were awarded the Nobel Prize in physics in 1956.

In this historical example, one can see how university science, pursued as basic research (basic knowledge) about nature (atomic nature) created the quantum theory of the atom (quantum mechanics). This occurred over the decades from 1900 to 1930. Key scientists in the discovery of the quantum nature of the atom were Max Planck, Ernest Rutherford, Niels Bohr, Albert Einstein, Werner Heisenberg, Ernst Schrodinger, Pascual Jordan, Wolfgang Pauli, Paul Dirac, and others. Next, an industrial researcher, Mervin Kelly, sought to replace the electron-vacuum-tube, enabled by the new science of quantum mechanics, which served as a knowledge base with which to explore a solid-state device (transistor). Kelly hired a team of young scientists, trained in the new quantum mechanical theory of solids (William Shockley, Stanley Morgan, John Bardeen and Walter Brattain). This team invented the transistor shortly after the end of the Second World War And another new age of electronics was born: a next-generation-technology for the electronics industry, based on the new science of quantum mechanics. We see here the importance of university research and industrial research, over time, to innovate the next generation of a technology.

History — Invention of the Integrated Circuit (IC) Semiconducting Chip

Next in the history of electronics, there occurred a change in the boundary of the technology system of the transistor — a transistor circuit on a silicon chip. Thus, while transistors were useful, a new technical problem had arisen in electronics. The complex circuits that engineers then could dream up with transistors required so many transistors that no one could physically wire them together by soldering transistors into printed circuit boards (the current technology of that time). A technical limit was on the horizon for transistorized electronics. How many transistors could be physically wired together? In 1958, this limitation was on the minds of two researchers: Jack Kilby at Texas Instruments, and Robert Noyce at Fairchild.

Jack St. Clair Kilby graduated as an electrical engineer and went to work at a firm named Centralab, located in Milwaukee. One evening he attended a lecture at the nearby Marquette University, given by John Bardeen, one of the inventors of the transistor. The new technology astonished Kilby, and he began reading everything he could find on the solid-state artifice. In 1951, Bell Labs announced the licenses, and Centralab purchased one for the $25,000 fee. They sent Kilby to Bell Labs for a five-day course in semiconductors. From 1951 to 1958, Kilby designed transistorized circuits for Centralab. As these circuits increased in complexity and as the number of transistors used increased, Kilby saw the need for a technical breakthrough beyond hand-wiring transistors together. But Centralab did not have the research capability which Kilby would need to tackle the problem.

In 1958, he sent his resume to several larger firms, including one to Willis Adcock of Texas Instruments. Adcock hired Kilby as TI was also worried about the integration problem of transistors. Kilby was assigned to a 'micro-module' research project at TI to research that problem, but he did not like their approach. One thing was clear to Kilby: since TI led in silicon technology, a good solution to transistor integration should use silicon technology — if it could. He had this idea: if transistors were made of silicon, could the other components of a circuit (such as resistors and capacitors) also be fabricated on silicon? Why not make the whole circuit on a silicon chip?

It was an inventive idea for technology then; carbon was used as the phenomenal material to fabricate resistors, while capacitors were made using plastics and metal foil. On July 24, 1958, Kilby wrote down what he called a 'monolithic idea' in his lab notebook: "The following circuit elements could be made on a single slice (of silicon): resistors, capacitors, distributed capacitors, transistor" (Reid, 1985). He entered rough sketches of how to make each component by properly arranging silicon material. Kilby showed his notebook to Adcock, who said that if he could demonstrate the idea worked, an R&D project to develop it would be authorized. Kilby carved out a resistor on one silicon chip and a capacitor on another and wired the two chips together. They worked, showing resistance and capacitance in the circuit. Adcock gave his approval to build a whole circuit on a chip and Kilby chose to build an oscillator circuit — one that could generate a sinusoidal signal.

On September 12, 1958, Kilby was ready. A group of TI executives assembled in the lab for a demonstration. The circuit on the chip was hooked to an oscilloscope, which would display the form of the signal from the circuit. Before the circuit was turned on, the Oscilloscope displayed a flat line indicating no signal output. Kilby switched on the new integrated circuit on the silicon chip and it worked — a wiggling line appeared on the screen, the sine-wave form of an oscillator circuit: "Then everybody broke into broad smiles. A new era in electronics had been born." (Reid, 1985)

Sometimes, a basic invention is made independently at about the same time by another inventive mind aware of the technological needs and opportunities. This happened with the invention of the semiconductor integrated circuit (IC) chip. That same summer in 1958, Robert Noyce, who was a physicist and then president of Fairchild Semiconductor, was also worrying about the problem of transistor integration. Fairchild was a new company which manufactured transistors, begun by Shockley, one of the inventors of the transistor, after he left AT&T. That following winter, on January 23, 1959, Noyce conceived a solution and wrote down in his lab notebook his idea: "It would be desirable to make multiple devices on a single piece of silicon, in order to be able to make interconnections between devices as part of the manufacturing process...." (Reid, 1985)

Both Texas Instruments and Fairchild Semiconductors filed for the patent on the IC chip. Years of legal argument followed about who had the right to the patent. In 1966, several semiconductor chip manufacturing firms met together, and TI and Fairchild agreed to grant licenses to each other and share in inventing the device. Later, Kilby and Noyce both received (in addition to many other prizes) the National Medal of Science for inventing the semiconductor IC chip.

The IC chip next became the key electronic component for progress in computers. Two kinds of IC chips were designed for computers: logic chips and memory chips. A logic chip contained all the circuits needed to do electronic computation in a computer, and was called a Central Processor (CPU) chip. A memory chip contained all the circuits needed to store information in memory, and was called a Dynamic Random Access

Memory (DRAM) chip. The two chips were the key components of a computer. Intel grew to dominate the global market for CPU chips. Other new American firms began the production of DRAM chips, but lost their business first to Japanese electronics firms, and these lost their business later to Samsung. To explain how this happened, we need to briefly review 'Moore's Law' and its role in planning IC chip progress.

Theory: Moore's Law as the Technology Roadmap for the Chip Industry

With the innovation of the IC chip, the new chip industry began planning progress in technology by increasing the density-of-transistors-on-a-chip by leaps and bounds — in factors of ten in chip density — for next-generations-of-technology in the transistor density of a chip. Deliberate planning for next-generations-of-technology in chips occurred in the global IC chip industry. Progress in chips occurred in plans for MSI (middle scale integration), LSI (large scale integration), VLSI (very large scale integration) and ULSI (ultra large scale integration). This international industrial planning happened because everybody in the industry knew about a technology roadmap for progress in IC chip technology. This was because back in 1965, Gordon Moore had told everybody.

Gordon Moore was the director of the R&D Lab of Fairchild Semiconductor. Moore anticipated continued progress in IC chips: "The complexity (of IC circuits) has increased at a rate of roughly a factor of two per year. Certainly over the short term this rate can be expected to continue, if not to increase. Over the longer term, the rate of increase is a bit more uncertain, although there is no reason to believe the rate will not remain nearly constant for at least 10 years (doubling each year)" (Moore, 1965). And so it happened, over the next twenty years, as shown in Figure 11.3.

Although Moore's law was not a 'law of nature', it worked as a '*technology roadmap*' in the semiconductor industry, and those working in the industry planned to make it happen.

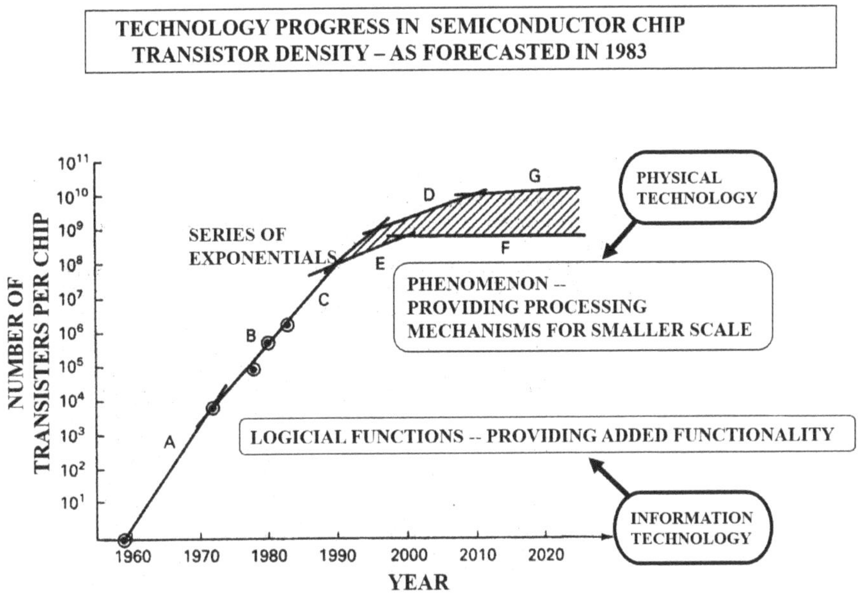

Figure 11.3. Moore's Law as a 1983 forecast of progress in semiconductor chips

Moore's law was used by the global industry as a guide to long-term technology planning. One notes that in Figure 10.1, the scale is semi-log, so the progress from 1960 to 1985 had been in two exponentials curves (graphed as straight-lines in semi-log charts). Ever smaller transistors were fabricated with new processing technologies. This enabled the rapid increase in the number of transistors on a chip. It was progress in chip-processing technologies which made possible the progress in IC chips. The increasing density of the transistors on a chip made possible the chip's vastly improved logical functions of processing information. *Information processing by IC chips is the 'function' of the chip technology; while transistor configurations of the chip are the form (morphology) of the chip technology — function and form in technology.*

The limit to Moore's law would occur when the physical features, in the constructing of a transistor out of silicon atoms, would get down to nearly the size of a silicon atom itself.

Transistors cannot be made smaller than their components, silicon atoms. The technology forecast even back in 1985 was for continuing exponential progress until after 2010. Then the 'natural limit' of progress in the technology would have been reached as the chip-making-technology became advanced enough to produce features nearly the size of a silicon atom. The natural limit is the transistor feature size, and the smallest possible transistor feature was predicted to be about the size of a hundred silicon atoms (or 10 nanometres). In 2012, chips were being produced at about the 20 nanometre scale and were expected to approach the 10 nanometre size around 2020. Thereafter, there would be no more Moore's law. The amazing thing is that the technology foresight by the chip industry in 1985 has been reasonably accurate. They knew that silicon atoms are finite in size. Nature eventually limits all technological progress and physical technologies. *This is because all technologies are ways to manipulate (alter, use) nature, and limitations on how much one can manipulate nature limits technological progress.*

Moore's law thus provided a technology roadmap to the world semiconductor industry. But which companies had the financial power and commitment to realize this goal? This would determine who captured the market. In 1981, at the technology discontinuity of the 64K Dynamic Random Access Memory (DRAM) chip, Japanese firms drove U.S. firms from the DRAM market. Korean firms drove Japanese firms from dominance in the DRAM market in 2000 and drove Taiwanese firms away in 2008.

History: Japan Innovates Transistors in Consumer Electronics

The transistor and IC chips were amazing inventions by Americans. But the focus then turns to American commercial activities in consumer electronics. American consumer electronics firms did not innovate (commercialize) the transistor for U.S. consumer products. *Instead the American invention of the transistor was rapidly used by American firms only for business and military markets, but not for consumer markets.* This left an opening for another nation's industry to enter the global market of consumer electronics, and Japan took this opportunity.

During the 1960s and 1970s, new US small businesses started to produce transistors and then IC chips. Yet it was the Japanese electronics firms in the 1980s, which had led the world in consumer electronics, which were the first to use transistors. Sony and other Japanese firms innovated the transistorized pocket radio, transistorized television sets, transistorized hi-fi, and home VCR's. *This technological substitution of transistors for tubes (technology discontinuity) in consumer electronics products provided the commercial opportunity for the Japanese electronics firms to grow, when U.S. consumer electronics firms lagged and were slow to make this change.*

The Japanese industry's extraordinary success in transistorized consumer electronics encouraged the Japanese government to assist the industry to go from transistors to integrated circuits, (IC chips). *The government wanted to assist Japanese industry to technologically advance from transistors to semiconductor IC technology (another technology discontinuity).*

History: Japan Innovates in DRAM Memory Chips

By the end of the 1960s, progress in IC chip technology had jumped from putting hundreds of transistors on a chip to putting thousands of transistors on a chip, called Middle Scale Integration (MSI). The next technology step after MSI was to put tens of thousands of transistors on a chip, called Large Scale Integration (LSI). In 1970, the Japanese Ministry of International Trade and Industry (MITI) sponsored a national program for LSI next-generation technology. The LSI scale for which the MITI project aimed was intended to bring Japanese electronics firms into competitiveness with U.S. firms. This decision by a nation to enter a new technology at the next-generation of the technology (rather than the current generation) was at the time an unusual decision for a developing economy. *This was a government-university cooperative research project to advance national technological competitiveness.*

MITI provided the equivalent of $10 million in research grants to Japanese electronics firms from 1970 to 1975. Some of the grants went to electronics firms producing transistors and consumer electronics devices, and some to firms making chip-production equipment. This was a government-industry research cooperation project. Japanese firms engaged in the project registered about 1000 patents on chip technologies. Their focus was on the memory chip using LSI technology capability, a 16K memory chip for computers, with a 16-thousand-bit storage capacity for dynamic random access memory (DRAM). This was smart Science & Technology policy.

Next, in 1976, this 16K memory chip was innovated by U.S. firms and by Japanese firms (assisted by the MITI-sponsored national LSI project). The demand for the 16K chip was so large by 1979 that U.S. firms could not satisfy the demand; Japanese firms started exporting to the U.S., gaining a 20% share of the U.S. market. U.S. computer manufacturers were surprised to find that the Japanese 16K memory chips were of higher production quality than memory chips produced in the U.S. About 20% of U.S. produced chips were defective, whereas only 10% of Japanese produced chips were defective. By 1980, the Japanese electronics firms were fully competitive with the U.S. in memory chip technology and more advanced in manufacturing quality.

Then in 1980, a next-generation of memory chip was aimed at in the 64K DRAM memory chip. Making it would require scaling down the size of transistors from LSI processing technology to Very Large Scale Integration (VLSI) technology. In 1981, the race was on between the U.S.'s and Japan's chip makers to be the first to innovate (commercialize) the 64K memory chip. The first to the market would likely grab the majority of the market; Japanese firms won out in November 1981.

U.S. firms had designed a 64K chip with some refined features over the 16K chip design to produce a smaller sized chip, but this redesign slowed down the timing of their product introduction. In contrast, Japanese firms took a standard 16K chip design (innovated earlier by the U.S. firm Mostek) and simply enlarged the design to the 64K level. They understood the importance of being the first to market. This resulted in a larger sized chip, but at the time, this made little difference to the computer customer. What mattered to the customer was speed (who produced the chip first) and quality (fewest defective chips). In September of 1982, when U.S. chip makers introduced their 64K memory chip, it was too late. Japanese DRAM chips already had 76% of the market. U.S. memory chip-makers died from lack of cash flow in selling their chips. At this technology discontinuity, Japanese industry better understood the conditions for success in global competitiveness than did their U.S. counterparts — Schumpeterian economic discontinuity.

The American semiconductor memory chip firms never recovered. The U.S. chip industry stopped producing memory chips (with a few exceptions). When chip density increased by an order of magnitude, the cost of building a chip processing facility to produce that transistor density also increased by about an order of magnitude. Unless a chip producer had a substantial market share, there was not enough cash flow to finance the investment for the next-generation production facility. Furthermore, each new facility to produce a new next-generation IC chip cost much more than the last factory. If profits were not coming in from the sales of the current chip, a small company would have no funds to pay for building the next-generation chip. U.S. chip-producing firms that lost the market share dropped out of the race for memory chip markets for good.

Technology Discontinuity and Schumpeterian Economic Discontinuity

Technology progress in IC chips had been anticipated and planned according to Moore's law. But which firms benefited from this knowledge? Only firms large enough and willing to make large investments of necessary capital could take practical advantage of the knowledge about the future. The move from technology discontinuity to economic discontinuity is what Schumpeter had called 'creative destruction' in the dynamics of technology and economy. *In creative destruction it is planning, commitment, and capital which make a difference in capturing the benefit of knowledge about a technology's future. A new superior industrial technology will replace a prior inferior industrial technology.*

In **Figure 11.4**, we describe the theory of the industrial life cycle curve for the DRAM industry as it occurred in history from 1961 to 1981.

Due to the planned technological progress by the semiconductor IC industry in the 1960s and 1970s, the product performance for the DRAM memory chips did not plateau, but continued to jump in 1970 and 1980 due to the technology discontinuities created by the next-generation products of 16K

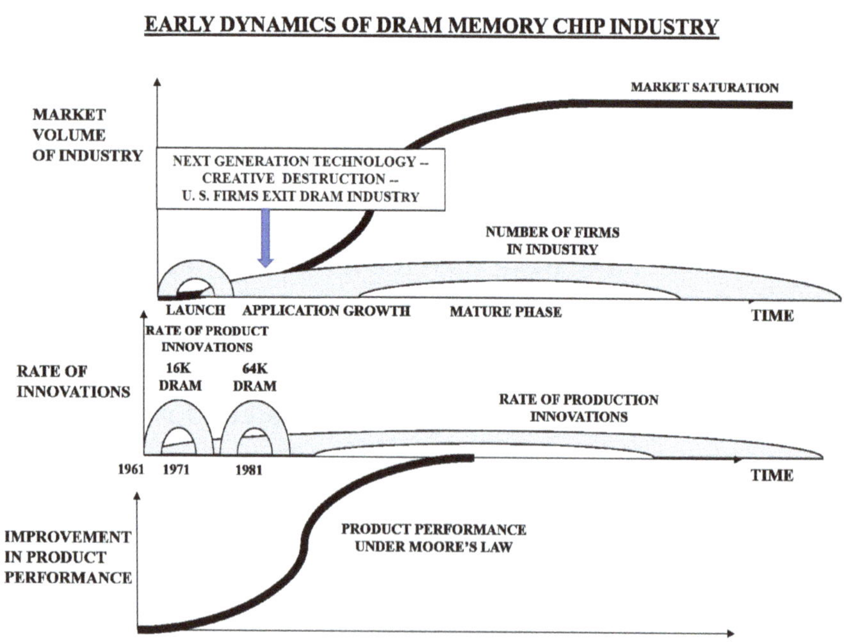

Figure 11.4. Dynamics of DRAM IC chip industry.

and 64K DRAM chips. But the market capture of the 64K DRAM chip by Japanese firms in 1981 (eight months before the U.S. firms introduced their 64K chip) had put the U.S. firms in jeopardy. Without an immediate cash flow from their 64K chips, the U.S. memory chip firms could not afford to continue in business (and compete for the next generation of DRAM chips, or the 264K DRAM chips). The market capture of the memory chip industry by the Japanese firms moved the industry from the U.S. to Japan, which can be seen as an example of Schumpeterian creative destruction.

In the case of IC chips, we now know that additional concepts must be added to innovation theory. The original key ideas of an industrial life cycle were:

(1) *The rate of innovation in the core technology of the product,*
(2) *The rate of innovation in the core technologies of its production,*
(3) *Number of firms entering and exiting the industry.*

Now in the memory-chip industry, there are the additional key ideas of:

(4) *Next-generation technologies,* and
(5) *Creative destruction.*

Next Generation Technology and Market Capture

To show the industrial life cycle for IC chips, one needs to add technology discontinuities in the form of next-generation transistor densities, as shown in Figure 11.5.

The chip industry has a history featuring multiple next-generation technology products, which makes it very different from an industry with only a single radical innovation. It did not develop a single product until peak performance was achieved, but instead developed several generations of products, with each generation of products performing dramatically better than the last. Production innovation is required in each

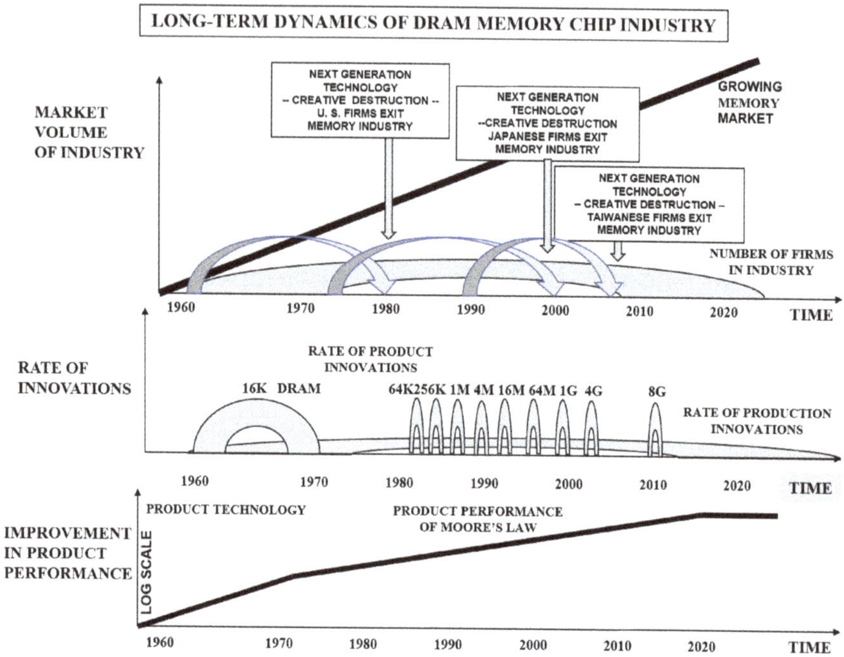

Figure 11.5. Dynamics of DRAM IC chip industry, with creative destructions.

generation to produce new products. The creative destruction of an industry can occur due to a technology discontinuity.

Product performance did not peak in the early days of the chip industry, but instead continued to improve due to next-generation chips. The next-generation chips were enabled by the next generation of production technologies. It was these technology discontinuities in production techniques which enabled ever-smaller transistors to be produced, which provided greater densities for increased memory capacity. But when the size of a transistor shrinks down towards the size of an atom, then progress in transistor chip density must end since transistors are made of silicon atoms and cannot be any smaller than that.

As the product performance was continuing to exponentially improve, the applications of the products expanded, and the market for memory chips did not saturate. Market growth continued to increase over the long term of the industry from 1960 through 2020.

But in each generation of new technology, the cost of factories increased by an order of magnitude. Progress in IC technologies came at ever large capital costs. This is why speed-to-market was so important for every generation of new technology. The first to exploit a technology discontinuity in the market captured the sales revenue, enabling it to continue R&D investment and factory investment for the next-generation technology. If a firm fell behind, sales ended and funds for continued investments stopped. The firm went out of the business of memory chips. But the Schumpeterian 'creative destruction' happened not just to an individual firm but to a nation's firms — national industrial destruction. Why? The answer lies in the capital structure of a nation for industrial development.

Innovation and Capital Structures

The capital structures for the U.S., Japan, Korea, and Taiwan were all different at the time of creative destruction in the memory-chip industry. At the time of the 1981–82 creative destruction of the U.S. memory chip industry, the U.S. interest rates (in the U.S. capital structure) on commercial loans ranged from 6% to 9%, in comparison to 3% interest, which the people of Japan were receiving from savings accounts in the Japanese Postal Office. These savings were available to the keiretsu as a source of capital and this meant that capital costs for the Japanese chip industry were one-third to one-half less than capital costs for the U.S. chip industry, putting the U.S. firms at a serious disadvantage in manufacturing investment.

A Japanese keiretsu was a group of interlocking companies controlled by a core bank. It could facilitate innovation as companies could technically provide mutual assistance within their keiretsu, and the core bank, which could provide for the commercialization of new technologies, could finance them. This form developed early in the industrialization of Japan and was restored during Japan's post-war reindustrialization in the 1950s. By 1980, Japanese electronics firms produced not only chips, but also consumer electronic products, generating revenue from both components and consumer products. Earlier in 1971, they had been assisted by the Japanese governmental agency MITI to acquire the semiconductor technology of Large Scale Integration, producing 16K memory chips by 1976. These electronic firms were each a part of a keiretsu and were financed by their keiretsu bank to build 16K chip production capacity in 1976 and then, significantly, 64K chip production capacity in 1981.

In contrast, the U.S. semiconducting chip firms were small single business firms, and did not produce electronic consumer goods. They had no cash flow from any product other than their semiconductor IC chips. They belonged to no corporate group with a bank at the centre, such as a keiretsu. Consequently, the better-financed, keiretsu-organized Japanese electronics firms grabbed the 64K DRAM market first. U.S. firms did not have the cash flow to continue to invest in the next generations of chips. All the U.S. firms devoted solely to memory chips went bankrupt. The only U.S. semiconducting IC firms to survive were those that produced logic chips (e.g. Intel, Motorola, Texas Instruments). The Schumpeterian creative destruction of the U.S. national memory chip business was due to the flaws of capital structure and corporate organization in the U.S., which together had not facilitated investments in next-generation chip product.

In 2000, a Schumpeterian creative destruction occurred in the Japanese memory chip industry when Samsung captured the global DRAM memory chip market from the Japanese electronics firms.

Respective capital structures have played an important role in those technology discontinuities. Samsung was a commercial group called a 'chaebol'. The chaebols were groups of interlocking businesses set up or acquired by an individual, sort of a 'family empire'. Chaebol companies owned stock in each other, and were able to finance each other's operations, because they were ultimately under control of the founding family. In the early industrialization of Korea after 1961, government policy encouraged the chaebol to acquire foreign technology to build new technology-based businesses for export. Then, the government selected some chaebols to undertake major technology acquisition to support project developments by guaranteeing their foreign loans. In the Samsung Group, cash flow from other Samsung companies and foreign loans provided the capital for Samsung Electronics to succeed in the development and production of a series of technology discontinuities from the 64K DRAM chip to the 4G DRAM chip. In 1996–97 during the Asian financial crisis, South Korea was hit hard. The Samsung group had large foreign debts, but it used the crisis to restructure the group and attain a sound financial basis. Afterwards, with large revenues from the 4G and 8G DRAM and NAND flash-drive memory discontinuities, Samsung dominated memory chip production. The Korean Schumpeterian creative destruction first drove Japanese, and then Taiwanese competitors out of the memory chip business.

The South Korean government had encouraged Korean industry to pursue advanced technology for export products — research cooperation between government and industry.

Summary

In trying to study Samsung's amazing success in the global market of memory chips in the twenty-first century, we used several important concepts:

— Technological Innovation = Invention + Commercialization + Market-capture,
— Importance of Capital Structures to Innovation,
— A revised Industrial Life Cycle for technology discontinuities and creative destructions,
— Technology Strategy for NGT Concurrent Development Programs in Speed-To-Market.

In the concept of technological innovation, the long-term history of an industry should include the idea of market capture along with the ideas of invention and commercialization. Technological innovation depends on 'invention' and 'commercialization' to begin the industry, but the success and domination of firms does not solely depend on these two ideas for survival. There is an additional component of 'market capture' that explains competition within an industry over a long time and that creates technical competitiveness.

Capital structure has an important relation to innovation. It is well known that the structure of venture capital is essential to radical innovation in its early stages of commercialization. But what has not been so well understood is the relationship of a capital structure to the later part of innovation, market capture. In the case of the IC memory industry, the capital structures of keiretsu in Japan and chaebols in Korea played important roles in sustaining the financing of innovation through several technology discontinuities.

Because of the importance of capital to commercialization, a Next Generation Technology (NGT) discontinuity in the core technologies of a product offers opportunities to disrupt and restructure an industry — creative destruction. To create a new order, we need to disrupt the equilibrium of the current state of play.

A series of NGT discontinuities can be technically planned in a technology roadmap, followed by NGT development programs which are put in place to speed up the pace through the technology discontinuities. But it takes a clear NGT technology roadmap and a strong CEO leadership to mastermind capital-intensive capabilities. The management skills to successfully carry out the development of NGT products ahead of time are also required.

Success in global completion requires innovation capability and management leadership with access to large capital. These three factors in innovation — technology strategy, leadership and capital — must together be implemented as a long-term strategy to capture a global market. *This is the basis of a technology or Schumpeterian discontinuity, and solving it requires research cooperation between government, industry, and university.*

Chapter 12

Summary: Cooperative Science & Technology Policy

Introduction

We have seen that at the national level, innovation requires research cooperation from the three research sectors in a nation: government, university and industry. This is necessary because the innovation process goes from (1) the scientific study of nature to (2) the technology used to manipulate it and then to (3) product development to ensure (4) sales of high-tech products/services as (5) utility. *This process requires connections between science, technology, and economy — and thus cooperation between universities, government, and industry.* Universities perform the most science research and are funded by government research agencies. Industry performs the most technological research, developing and designing the new technology into new high-tech products and services that drive the economy.

Science and technology interact. Science provides the knowledge base and research techniques for technology development; technological progress improves scientific instrumentation. Industries perform technological development, and technology provides and creates industries with productive competency. Interactive science and technological progress is necessary for a nation to be competitive, and this requires research cooperation between universities, industries and the government.

Shown in Figure 12.1 are the four forms of effective cooperation : Science Cooperation, R&D Cooperation, Development Cooperation, and Targeted-Basic-Research Cooperation.

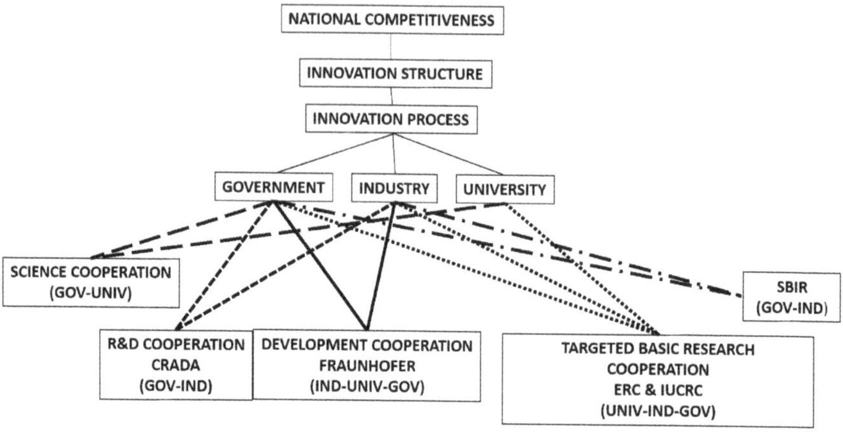

Figure 12.1. Types of cooperative research programs created by government S&T policy.

The three research sectors of a nation (government, university, industry) all require the *capability to innovate*. There are four effective institutional ways to facilitate research cooperation across the sectors.

- First, there is *Science Cooperation* between government and industry, as exemplified by the U.S. NSF's support of university research.
- Second, there is *R&D Cooperation* between government and industry, as exemplified by the U.S. NIH's CRADA program.
- Third, there is *Development Cooperation* between industry, university, and government, as exemplified by the German Fraunhofer Society.
- Fourth, there is the *Targeted-Basic-Research Cooperation*, as exemplified by the US NSF ERC and IUCRC programs.

National Capability for Innovation

National innovation requires both a structure and a process. We know this strategy: universities are the primary source of scientific progress. Industries are the primary source of technological progress. Government is the primary source of science funding and encourages the development of radical innovation. Science and technology interact, driving progress in both areas. Research cooperation is important here because no nation can be economically successful without being competitive in both science and technology. Interacting together, science and technological progress pave the way for discovery and invention. Industry transforms invention into utility, using research & development activities, based upon scientific knowledge. This creates the structure for national innovation.

We also know this general pattern: nature is valuable to society when nature can be manipulated in ways that make it useful to humans. Innovation consists of the generation of knowledge as the transformation of nature into something useful — by applying science and technology to it, and subsequently embedding it in an economically viable product or service. This forms the process for national innovation. What is particularly striking about this is that two different perspectives are involved: (1) a technical perspective about what is important in the world of innovation and (2) a financial perspective emphasizing how much profit can be made from innovation. The innovation process requires two perspective foci — on the world as material (technical) and on the world as financial (money). Entrepreneurship is the basic management style for innovation, as entrepreneurs emphasize both new technology and profits.

The scientific knowledge of nature provides a 'knowledge base' upon which we can build to invent new technologies. Usually engineers or scientists or other technical personnel are the ones who invent and improve technologies. Engineers use technical knowledge to design products, processes, services and software. Commercialization is the act of using technology to design products, services and processes and producing these for sale. A business competes by selling high-tech products and/or services, generating income and therefore profits.

Science Cooperation between Government & University

For universities to have *depth in scientific research capability*, the government should establish a special government agency to fund disciplinary research projects in universities. In the U.S., this is provided by the National Science Foundation. In Germany, this is provided by the Max Planck Society. Government funding agencies should use 'peer review' in selecting research proposals for funding to ensure scientific quality and to avoid political influence in science. The societal benefits of disciplinary research take place over a long period of time and can be unpredictable.

Targeted-Basic-Research Cooperation

To have university basic research also contribute to technological innovation in a short time period, complementary funding programs for research centers at universities which are multidisciplinary and industrially-cooperative are useful. The ERCs and IUCRCs funded by the U.S. NSF are examples of such centers.

R&D Cooperation between Government & Industry

Government research-funding agencies should have policies to transfer knowledge and intellectual property from government to industry. In the U.S, a cooperative research agreement between a US government agency and a company is called a Cooperative Research and Development Agreement (CRADA). For example, an NIH CRADA provided opportunities for NIH government researchers to work with industrial and university researchers toward common research goals.

Development Cooperation between University & Industry & Government

In national innovation, businesses generate revenues directly from the sale of products and services that are derived directly from engineering research, but not from science. Innovation intermediaries, such as the German Fraunhofer Institutes, have played an important role in German national economic success. Fraunhofer Institutes directly assist German businesses, large and small, to undertake both research and development (R&D) commercially and successfully to innovate and design new products, processes, or services — all of which contribute directly to driving sales and generating revenue.

Small Business Research Cooperation

In the United States, the German Fraunhofer model was never copied. Instead in 1982, the U.S. established a new program for the government funding of small businesses called Small Business Innovation Research (SBIR). All government research funding agencies were required to spend 2.5% of their research budgets on research grants to small businesses. In 2017, the percentage funding was subsequently raised to 3.2%. Federal agencies participating in the program include the Department of Defense, the National Institutes of Health, the National Science Foundation, and the Department of Agriculture.

Summary

Because of the different roles and capabilities of the three research sectors for national innovation, the governmental science & technology policy of a nation should:

(1) fund university research to produce Ph.D. graduates as a knowledge resource for the nation, and
(2) facilitate research cooperation between the research sectors of government, industries and universities.

References

Abernathy, W.J., and K.B. Clark. 1985. "Mapping the Winds of Creative Destruction", *Research Policy*, Vol. 14., No. 1, pp. 2–22.

Abelson, P.H. 1996. "Pharmaceuticals Based on Biotechonogy", *Science*, Vol. 273, No. 5276, p. 719.

Abramson, H.N., J. Encarnacao, P.P. Reid, and E. Schmoch (Editors). 1997. *Technology Transfer Systems in the United States and Germany: Binational Panel*. National Academy Press, Washington D.C.

Ayres, R.U. 1989. "The Future of Technological Forecasting", *Technology Forecasting and Social Change*, Vol. 36, No. 1-2, pp. 49–60.

Betz, F. 2010. *Managing Technological Innovation, 3rd Ed.* John Wiley, New York.

Betz, F. 2015. *Strategic Thinking – Vision, Competence, and Integrity.* Emerald Group Publishing, Cambridge, Massachusetts.

Betz, F., Y. Choi, and W. Min. 2012. "Innovation, Capital and Market Capture: The Case of Samsung", *PICMET 2012 Conference*, Vancouver, Canada.

Betz, F., Y. Choi, W. Min, W.K. Park, and D.W. Shin. 2014. *Innovation and Global Competition: The Case of Korea*. Nova Science Publishers, New York.

Betz, F., W. Min and D.W. Shin. 2014. "Universities and Entrepreneurship in Asia: The Case of Nano." *Journal of the Knowledge Economy,* Vol. 5, No. 4, pp. 803–819.

Biotechnology Process Engineering Center. n.d. Retrieved from http://web.mit.edu/bpec/.

Bush, V. 1945. *Science: The Endless Frontier*. United States Government Printing Office, Washington, D.C. Retrieved from https://www.nsf.gov/od/lpa/nsf50/vbush1945.htm

Carayannis, E.G., and D.F.J. Campbell. 2009. "'Mode 3' and 'Quadruple Helix': Toward a 21st Century Radical Innovation Ecosystem", *International Journal of Technology Management*, Vol. 46, No. 3-4, pp. 201–234.

Carayannis, E.G., and D.F.J. Campbell. 2010. "Triple Helix, Quadruple Helix and Quintile Helix and How Do Knowledge, Innovation, and Environment Relate to Each Other?" *International Journal of Social Ecology and Sustainable Development (IJSESD)*, Vol. 1, No. 1, pp. 41–69.

Carayannis, E.G., and D.F.J. Campbell, 2012. *Mode 3 Knowledge Production in Quadruple Helix Innovation Systems*. Springer Briefs in Business, New York.

Cordis.europa. 2007. Retrieved from https://cordis.europa.eu/programme/rcn/837_en.html.

De la Merced, M.J. 2017. "Gilead to Buy Kite, Maker of Cancer Treatments, for $11.9 Billion", The New York Times, August 29. Retrieved from https://www.nytimes.com/2017/08/28/business/dealbook/gilead-kite-gene-therapy.html.

Grady, D. 2017. "FDA Approves First Gene-Altering Leukemia Treatment, Costing $475,000", The New York Times, August 31. Retrieved from https://www.nytimes.com/2017/08/30/health/gene-therapy-cancer.html.

Grady, D. and A. Pollack. 2016. "What Is Immunotherapy? The Basics on These Cancer Treatments", The New York Times, July 30. Retrieved from http://www.nytimes.com/2016/07/31/health/what-is-immunotherapy-cancer-treatment.html.

Gwynne, P. 1993. "Directing Technology in Asia's 'Dragons'", *Research-Technology Management*, Vol. 36, No. 2, pp. 12–15.

Hauben, Michael. 1993. History of ARPANET. Retrieved from http://pages.infinit.net/jbcoco/Arpa-Arpanet-Internet.pdf.

IMS. n.d. Retrieved from www.imscenter.net

Jetter, A.J. 2010. "Aachen University Fraunhofer Institute." Private Communication.

Jetter, A.J. and S. Munkongsujarit,. 2013. "The Impact of Social Capital on the Outcomes of Innovation Intermediation Process in the IT-Driven Services", *PICMET 2014 Conference*, San Jose, California, USA.

Kahn, B. 1990. "The Software Patent Crisis." *Technology Review*, April, pp. 53–58.

Letzkowitz, H., and Leydesdorff, L. 2000. "The Dynamics of Innovation: From National Systems and "Mode 2" to a Triple Helix of University–Industry–Government Relations", *Research Policy*, Vol. 29, No. 2, pp. 109–123.

Markoff, J. 1999. "An Internet Pioneer Ponders the Next Revolution", New York Times, December 20. Retrieved from http://movies2.nytimes.com/1999/12/12/technology/122099outlook-bobb.html

Marquis, D.G. 1969. "The Anatomy of Successful Innovations", *Innovation*, Vol. 1, pp. 28–37. (Reprinted in *Readings in the Management of Innovation*, M.L. Tushman and W.L. Morre, (eds.), Pitman, Marshfield, Massachusetts, 1982.)

Meyer-Stamer, J. and F. Wältring. 2000. "Behind the Myth of the Mittelstand Economy, The Institutional Environment Supporting SME Enterprises in Germany", Institute for Development and Peace at the Gerhard-Mercator-University Duisburg. Retrieved from http://edoc.vifapol.de/opus/volltexte/2013/4560/pdf/report46.pdf.

Mitchell, T. 2010. *Samsung Electronics*. John Wiley & Sons, Hoboken, New Jersey.

Morone, J.G. 1993. "Technology and Competitive Advantage—the Role of General Management," *Research-Technology Management*, Vol. 36, No. 2, pp. 16–25.

NIH. n.d. Retrieved from http://www.nih.gov.

NSF. n.d. Retrieved from http://www.nsf.gov.

Nelson, R. 1993. *National Innovation Systems: A Comparative Analysis*. Oxford University Press, Oxford, UK.

NSF. 2016. *Science Indicators*. National Science Foundation. Retrieved from https://www.nsf.gov/statistics/2016

Olby, R. 1974. *The Path to the Double Helix,* Macmillan, London; University of Washington Press, Seattle, WA.

Pollack, A. 2015. "Riding High, Biotech Firms Remain Wary", New York Times, January 18. Retrieved from https://www.nytimes.com/2015/01/19/technology/riding-high-biotech-firms-remain-wary.html

Pollack, A. 2016. "Setting the Body's 'Serial Killers' Loose on Cancer", New York Times, August 1. Retrieved from http://www.nytimes.com/2016/08/02/health/cancer-cell-therapy-immune-system.html?rref=collection%2Fseriescollection%2Fimmunotherapy&action=click&contentCollection=health&region=rank&module=package&version=highlights&contentPlacement=1&pgtype=collection

Porter, M. 1990. *Competitive Advantage of Nations.* The Free Press, New York.

Portugal, F.H., and J.S. Cohen. 1977. *A Century of DNA*, MIT Press, Cambridge, Massachusetts.

Reid, T.T. 1985. "The Chip", *Science 85*, February pp 32–41.

Richtel, M. and A. Pollack. 2016. "Harnessing the US Taxpayer to Fight Cancer and Make Profits", New York Times, December 19. Retrieved from http://www.nytimes.com/2016/12/19/health/harnessing-the-us-taxpayer-to-fight-cancer-and-make-profits.html?_r=0

Robaschik, F. 2013. "Seminar zum deutschen Mittelstand", Deutschen Botschaft in der Republik Korea, March 5. Retrieved from http://www.seoul.diplo.de/Vertretung/seoul/de/__events/mittelstand-v.html?archive=3765102

U.S. Department of Defense. n.d. Small Business Innovation Research. Retrieved from http://www.acq.osd.mil/osbp/sbir/about/index.shtml

U.S. Small Business Administration. n.d. SBIR-Participating Agencies, https://www.sbir.gov/about/about-sbir#three

Teraski, M. 2016. "A Finer Look at a Fine Cellular Meshwork", *Science*, Vol. 354, No. 6311, pp. 415–416.

Thayer, A. 1996. "Market, Investor Attitudes Challenge Developers of Biopharmaceuticals." *Chemical & Engineering News.* August 12, pp. 13-21.

Tibbets, R. 2009. "Reauthorizing SBIR: The Critical Importance Of SBIR And Small High Tech Firms In Stimulating And Strengthening The U.S. Economy", Annex 3 in The Role Of The SBIR And STTR Programs In Stimulating Innovation At Small High-Tech Businesses: Hearing Before The Subcommittee On Technology And Innovation Committee On Science And Technology. Retrieved from http://www.gpo.gov/fdsys/pkg/CHRG-111hhrg48735/pdf/CHRG-111hhrg48735.pdf.

Tubitak. n.d. Retrieved from http://www.tubitak.tr.

Wolff, M.F. 1983. "Mervin J. Kelly: Manger and Motivator", *IEEE Spectrum*, May, pp. 81–85.

Index

Lightning Source UK Ltd.
Milton Keynes UK
UKHW021146150119
335605UK00002B/49/P

9 789813 238695